Praise for
The Bigger Picture

'Although they will perhaps always remain mysterious, psychedelics have also always invited us – and our ancestors – to heed the teachings they have to offer, to reconsider our place in the universe, and to recognize the enigma of our own existence. *The Bigger Picture* adds substantial new information to our understanding of psychedelics and their potentially momentous significance to the world today. I highly recommend this carefully and convincingly argued, deeply thought-provoking and beautifully written book.'

GRAHAM HANCOCK

'*The Bigger Picture* is an entertaining, insightful, and timely book. Beiner draws on the latest psychedelic science to explain how what we're learning about these molecules can help us make sense of our social and political challenges in new ways, and he ties together research from a variety of different fields to present a compelling and nuanced argument as to why psychedelic science could change the world for the better.'

ROBIN CARHART-HARRIS, FOUNDER OF THE CENTRE FOR PSYCHEDELIC RESEARCH AT IMPERIAL COLLEGE LONDON, AND DIRECTOR OF PSYCHEDELICS DIVISION, NEUROSCAPE, UCSF

'Beiner is a wise and entertaining guide through the jungle of contemporary psychedelics, bringing much-needed critical thinking to the field'.

JULES EVANS, AUTHOR OF *PHILOSOPHY FOR LIFE* AND DIRECTOR OF THE CHALLENGING PSYCHEDELIC EXPERIENCES PROJECT

The
Bigger
Picture

The
Bigger
Picture

How **Psychedelics** Can Help Us
Make Sense of the World

Alexander Beiner

HAY HOUSE

Carlsbad, California • New York City
London • Sydney • New Delhi

Published in the United Kingdom by:
Hay House UK Ltd, The Sixth Floor, Watson House
54 Baker Street, London W1U 7BU
Tel: +44 (0)20 3927 7290; Fax: +44 (0)20 3927 7291
www.hayhouse.co.uk

Published in the United States of America by:
Hay House Inc., PO Box 5100, Carlsbad, CA 92018-5100
Tel: (1) 760 431 7695 or (800) 654 5126
Fax: (1) 760 431 6948 or (800) 650 5115; www.hayhouse.com

Published in Australia by:
Hay House Australia Ltd, 18/36 Ralph St, Alexandria NSW 2015
Tel: (61) 2 9669 4299; Fax: (61) 2 9669 4144; www.hayhouse.com.au

Published in India by:
Hay House Publishers India, Muskaan Complex,
Plot No.3, B-2, Vasant Kunj, New Delhi 110 070
Tel: (91) 11 4176 1620; Fax: (91) 11 4176 1630; www.hayhouse.co.in

Text © Alexander Beiner, 2023

The moral rights of the author have been asserted.

The information given in this book should not be treated as a substitute for
professional medical advice; always consult a medical practitioner. Any use
of information in this book is at the reader's discretion and risk. Neither
the author nor the publisher can be held responsible for any loss, claim
or damage arising out of the use, or misuse, of the suggestions made, the
failure to take medical advice or for any material on third-party websites.

A catalogue record for this book is available from the British Library.

Tradepaper ISBN: 978-1-78817-915-7
E-book ISBN: 978-1-78817-916-4
Audiobook ISBN: 978-1-78817-917-1

MIX
Paper from
responsible sources
FSC® C013056

Printed and bound in Great Britain by
TJ Books Limited, Padstow, Cornwall

Contents

Introduction:
In and Through

They stand over my bed. I'm blindfolded, flat on my back. My heart is pounding. An electrode helmet on my scalp measures my brain waves and spacey music fills my ears. My mind is focused on the ache in my left arm, where a tube connects me to a dose of the world's most powerful psychedelic molecule.

Any moment now, a scientist will press the button on that pump. The drug will shoot into my blood and reality will flip upside down. I'll plunge into an experience so strange and profound that nobody, scientist or mystic, understands it. They'll keep me there for 40 minutes as I embark on a voyage I've spent months preparing for.

Knowing all this, I do my best to stay calm as I wait for the injection. I focus on my breathing. It's hard. My throat is dry and my mind is racing. I try to distract myself, thinking of mountains and calm streams – but now, with a cold sting, the drug comes in, and I gasp and have a final thought.

Why did I sign up for this?

It took me a long time to answer that question. To begin, it's easier to explain what I'd signed up for. It was autumn 2021 and I was participating in a research trial at Imperial College London, investigating the effects of the powerful psychedelic drug dimethyltryptamine (DMT) on healthy volunteers. The experience usually lasts 10 minutes. This trial explored what happened in our brains and our inner worlds when we were dosed continuously for 40 minutes.

Lying on that bed surrounded by scientists, I was about to embark on a voyage that would change me forever. And while what I discovered is woven through the book you're reading, it isn't ultimately about my personal experiences. It's about the collective voyage we're all on in this period of history. It's a time when our old certainties have collapsed and we face the very real threat of extinction; a time brimming with hope, danger, sorrow, and transformation. This book is about how what we're learning from psychedelic science and spirituality can help us find new ways to make sense of and come through the crisis of civilization.

From global warming to geopolitical instability and political polarization, we are in an age of chaos and fragmentation – not just of the world order, but of our hearts and souls. The mental-health crisis that is spurring a renaissance in psychedelic research is just one symptom of these much deeper crises, birthed by the gnawing hole of meaning and purpose at the heart of consumer culture.

Meeting these challenges requires a level of coordination, imagination, and compassion unlike anything we've managed before. But in a time when we're regularly disconnecting from one another in online worlds that let us live in completely different realities from our neighbors, we need new ways to come together and coordinate to tackle the biggest problems we face.

Throughout the book, I will explore how not just the insights we can gain, but the cognitive and emotional capacities we can develop through psychedelic experiences, can be applied directly to making sense of what's going on around us. I'll look at how human beings have been using psychedelic practices to move between different realities for thousands of years, and why we need to tap into those skills today. I'll explore how they can guide us deep into our own humanity and the essential wisdom hidden within it. Above all, I'll share how they can reconnect us to ourselves, to one another, and to the planet.

To find out how, I'll speak to leading scientists, philosophers, tricksters, and artists. I'll share emotional and cognitive techniques developed from the latest neuroscience and wisdom traditions to help us navigate complexity and look at the deeper metaphysical questions psychedelics raise. But first, I'll explain how and why it is that psychedelics are now entering mainstream awareness once again.

The Psychedelic Renaissance

If you haven't heard the term before, the word 'psychedelic' might conjure images of the late 1960s. The Summer of Love. Tie-dye. Timothy Leary telling people to 'turn on, tune in, and drop out' of society. Or perhaps your mind travels further back, to the dawn of civilization, when shamans communed with teaching plants to navigate the spirit world. Maybe it conjures images of the people around the world who are still using psychedelics as sacraments – from the use of iboga by the Bwiti of Gabon, or of the fly agaric by the Tungusic shamans of Siberia, to the ayahuasca ceremonies of the Shipibo-Conibo of Peru or the Santo Daime church in Brazil.

If you've clicked on an article about psychedelics in the last five years, chances are it wasn't about counterculture or indigenous use, but about psychedelics' potential to treat mental-health disorders. A new wave of research, investment, and changing cultural perceptions has sparked what's been called 'the psychedelic renaissance,' with billions of dollars now being poured into psychedelics in the hope that they may offer us a way out of a growing health crisis that sees close to a billion people struggling with some form of mental illness. Someone takes their own life every 40 seconds. A sense of dislocation, anxiety, and loneliness pervades modern life, and everyone's looking for a solution.

The Imperial College Psychedelic Research Group, who were running the study I participated in, have been the source of many of these headlines; they have overseen multiple studies investigating the therapeutic and neuroscientific effects of psychedelic drugs. Over the last decade, labs like this have been reexamining how psychedelic molecules like psilocybin (the active ingredient in magic mushrooms) can treat some of the most widespread mental-health conditions, including depression, anxiety, eating disorders, and obsessive-compulsive disorder (OCD). The results have been so promising that psilocybin has been given 'breakthrough therapy' status for treating depression by the FDA in the USA, while 'the love drug' MDMA is on the cusp of becoming a legal medicine for treating post-traumatic stress disorder (PTSD).

The study I participated in was different to the ones that normally make the headlines. It was non-therapeutic, designed to investigate what exactly is going on in the brain and the mind when we take DMT. Of all the psychedelic compounds, DMT has a particularly metaphysical mystique. That's partly due to the weird things that happen when human beings ingest DMT. Normally, the drug is vaporized in a small pipe and the experience lasts under 10 minutes. Despite the short duration, many people describe it as one of the strangest, most profound, and sometimes most terrifying experiences of their lives.

Some report traveling out of their bodies, visiting other intricate worlds populated by seemingly intelligent and independent entities. Sometimes these entities are friendly, sometimes aggressive, and always inexplicably weird. Many people report life-changing personal and metaphysical insights. DMT can bring us into a deep sense of connection with wider reality, and many say their experience felt 'realer than real.'

The study in which I took part was the first to investigate what happens when you extend that experience by a factor of four, and due to its highly experimental nature, only psychologically healthy (or healthy enough) volunteers with previous psychedelic experience were recruited. Even so, by the time I signed up, the researchers had decided to change the study to include four rather than five doses, as a few people had already dropped out because of the intensity of the trial.

That uncomfortable fact was one of the many fears going through my mind on that bed. I have a lot of experience with psychedelics, and as a podcaster and writer in the space, I'm up to speed on the latest research. Physically, I knew the experience was safe. But I was anxious about what I was going to encounter.

A wary caution is a common attitude among people experienced with psychedelics. However, it's often missing from all the media hype, which sometimes presents them as wonder drugs that can cure depression; this ignores just how complex these medicines are, and how little we really know about them.

What is certain is that psychedelics work very differently to normal psychiatric medicines. Psychiatrist Stanislav Grof, a pioneer in the field, has called them 'non-specific amplifiers.'[1] His theory is that they amplify what's already in our hearts and minds. They bring up memories, insights, and behavior patterns we aren't normally aware of, and often elicit a deep sense of meaning and connectedness.

However, you can't just take a psychedelic and hope you'll go through a profound healing process. In fact, they can be just as dangerous as they are healing. For psychedelics to be healing and effective, we need to experience them in a safe setting, within a strong therapeutic or ceremonial container that can hold us through the experience and help us make sense of our insights. When these conditions are met, psychedelics can do more than help us navigate our minds; they can elicit profound spiritual experiences.

In 2006, a Johns Hopkins study conducted by Roland Griffiths reported that more than half of participants ranked their psilocybin experience as one of the five most spiritually meaningful of their lives – right up there with marriage or the birth of a child. This 'mystical' encounter reported by so many is believed to be a key component of what makes psychedelic experiences so healing.[2]

My Story

A major impetus for this book is the fact that the mystical experiences I've had on psychedelics have forever changed my life and how I see reality. But during the study, I wasn't lying on that bed looking for a mystical experience, or at least I didn't think I was. I had a specific goal in mind – a personal experiment I'd be conducting throughout the four months of the trial, and a question that forms the main inquiry of the book you're reading: How can psychedelics help us find new solutions to the existential crises we're facing as a species?

This is a question I've been interested in since my first psychedelic experience at the age of 18, when I was with a group of friends at a park in the Dutch city of Maastricht. I grew up about a two-hour drive away, near Frankfurt, Germany. My father is German and my mother is from Northern Ireland. My father worked in the printing industry and my mother was a teacher at Frankfurt International School, which I attended from the ages of five to 18.

Many of us who've grown up in very multicultural environments, or with two nationalities, often have a strange relationship with the idea of 'home.' However, as I sat in that park in Holland experiencing mushrooms for the first time, I felt more at home than I'd ever felt. I had a calm and abiding feeling that I had reconnected to reality in a way that would change me forever.

Over the next few years, I devoured everything I could find on psychedelics. I became fascinated with the anthropology of shamanism, neuroscience, ethnobotany, and psychedelic philosophy. I spent hours listening to recordings of psychedelic pioneers like Terence McKenna, Ram Dass, Ann Shulgin, and countless others. I began to relate to psychedelics as sacred medicines and took them infrequently, carefully, and with great respect. I spent much of my time at university writing a novel about psychedelic culture and shamanism, *Beyond the Basin*, which I published just after I graduated. At around the same time, I began a daily meditation practice.

In 2009, as Occupy Wall Street was at its peak, I moved to London. I met my now-wife, Ashleigh Murphy-Beiner, who would go on to become a psychologist and a psychedelic researcher. We both trained as meditation instructors and set up Open Meditation, a school in London that taught non-religious mindfulness to people and companies.

In 2011, I spoke at a small psychedelics conference called Breaking Convention at the University of Kent. At the time, even the idea of talking about psychedelics in public felt risky. Studies about the topic were starting to come out, but the taboo was so strong that many speakers were doing the math to make sure any psychedelic experiences they recounted were more than seven years old, meaning they'd be outside the statute of limitations.

Despite our nervousness, it was a magical experience. Breaking Convention was one of the first times in the UK that people had gathered to talk about the academia, science, culture, and philosophy of psychedelics. Conversations that normally happened anonymously on Internet forums were now taking place in real life, with many people from different backgrounds coming out of the woodwork and tentatively speaking about something of profound importance. More than a decade later, Breaking Convention is one of the longest-running psychedelic conferences in Europe. I went on to join the organizing committee and eventually became one of the executive directors.

It was through Breaking Convention that I met David Fuller, a former BBC and Channel 4 journalist who was volunteering to help with our press outreach. In 2017, we founded an organization called Rebel Wisdom. We were both trying to make sense of the significant cultural shifts taking place after Brexit and the US election of Trump. Because we both had an interest in personal growth and transformative practices, we believed that meaningful sense-making requires a deeper level of inquiry into psychology, spirituality, and sociology.

Rebel Wisdom grew into a media and events platform with more than a quarter of a million subscribers. Over five years, we interviewed rebellious thinkers and doers from a variety of fields, and ran live events, courses, and retreats to help people apply practices like mindfulness, collective intelligence, and shadow work; all of these undertakings helped us make sense of what's going on in the media, culture, and politics. Above all, we and other organizations like us focused on what it will take to find a way to solve some of the complex problems we're grappling with collectively.

A key insight that began to emerge as we inquired into that question was that if we really want to adapt and thrive in the complex age we live in, we need to evolve not just what we think and do, but *how* we do it. This means staying curious and learning to flow with change and contradiction with an attitude of playfulness, compassion, discernment, and creativity. It also means learning how to see beyond our existing frames of reality to find new insights. Psychedelics can open our minds to this way of seeing the world, more reliably than anything else I'm aware of. And so, it was particularly poignant for me that the psychedelic renaissance started to explode into the mainstream as we explored these topics.

Between 2020 and 2022, psychedelic medicine received a gold rush of investment, with dozens of psychedelic pharma companies raising close to $2 billion in stock-market flotations. At the time of this writing, it's estimated that it will grow to be worth $8.3 billion globally by 2028.[5] The counterculture visions that fired up that first Breaking Convention started taking a backseat to boardroom negotiations. The positive clinical-trial results coming out of labs like Imperial and Johns Hopkins added a new kind of credibility to psychedelics, as did books like Michael Pollan's *How to Change Your Mind: The New Science of Psychedelics*.

The more recent climate around psychedelics is worlds away from what it was during that first Breaking Convention in 2011. The press stopped telling stories about people jumping out of windows high on acid, and instead started publishing glowing articles about the promise of psychedelic science. Some investors have even claimed that psychedelic pharma companies could solve the mental-health crisis for good.

What We're Missing

In 2020–2022, I started to cover this psychedelic gold rush journalistically. It started with a film I made called *The Rise of Psychedelic Capitalism*, exploring the nuances of what happens when a complex, transformative experience hits the realities of capitalist market forces. In the process, I noticed just how fast the narrative was changing, and how much was being lost in the race to medicalize psychedelics.

Psychedelics were being framed not as agents of social transformation, but as exciting new mental-health drugs. In many ways, this approach has been tremendously successful. Research is booming in psychedelic science. Celebrities like Will Smith now openly talk about their ayahuasca journeys. Psychedelic pioneer David Nutt, who leads the Imperial College Psychedelic Research Group, pointed out when we spoke that most of the resistance to legalization is now coming from more conservative clinicians within the medical establishment; a recent study suggested that almost 60 percent of the general public in the UK are in favor of reclassifying psilocybin.[3]

But despite its undeniable success, the commercialization of psychedelics has had unintended consequences. The experiences we have on psychedelics are highly influenced by our cultural expectations, as well as our set (our state of mind), setting (where we are), and dose (how much we take). Psychedelics alone don't necessarily lead to transformation; change is brought about when psychedelics are taken in combination with a practice like therapy or a religious ceremony.

When you change the dominant cultural narrative around these medicines, you change how people experience them. The risk is that big pharma, the well-being industry, and biomedical psychiatry create a narrative around what psychedelics are, who they are for, and how they are allowed to be used – which narrows their tremendous potential for social change. It also presents a real risk that, like many forms of therapy, psychedelics become accessible primarily to the wealthy.

I will return to the debate around psychedelic capitalism throughout the book, revealing just how complex it is. On one hand, all the investment pouring into psychedelics is creating mainstream acceptance and has the potential to help millions of people. On the other hand, it might come at the expense of the very message psychedelics often teach us: We need to change.

If you read enough glowing headlines about psychedelic science, or the many PR releases put out by psychedelic pharma companies, it can appear that we understand what needs to change to solve the 'mental-health crisis.' From the perspective of biomedical psychiatry, our depressions, anxieties, and addictions can be fixed with new drugs or therapies.

However, this is only one part of a much bigger story, and relies on an assumption that mental illness exists primarily *within* us. From another perspective, the mental-health crisis is a collective howl from the heart of consumer culture. A howl that warns of socioeconomic disparity, with children from underprivileged backgrounds four times more likely to suffer mental-health afflictions than wealthier peers. A howl that comes straight from the epidemic of our loneliness, and from skyrocketing rates of teenage mental illness driven by predatory social-media algorithms.

Clinical psychologist Lucy Johnstone has argued that placing the source of illness within the individual ignores the complex set of conditions that have led to their illness. She says we shouldn't be asking, 'What's wrong with you?' Instead, we should be asking, 'What happened to you?'[4] Johnstone represents a growing awareness among psychologists that it may be our environment, cultural values, and economic situation that drive our increasing rates of depression and anxiety, as well as our thoughts and biology.

The best chance we have of healing the mental-health crisis is by zooming out to look at the bigger picture and addressing it from the roots. This is no simple task, because it asks us to fundamentally change how we live, both individually and collectively. The early promise of the counterculture that sprang up in the 1960s and 70s was that psychedelics could do exactly this. They could offer a deeper understanding of how we've created our shared reality and give us the ability to make a new choice. They could decondition us from the cultural values that are making us so sick by reconnecting us not just to ourselves and one another, but to that which we have distorted in our hyper-individualistic, blindly consuming culture: the sacred.

The Voyage Ahead

Trying to find a way to renew this countercultural psychedelic vision for the times in which we live is the real reason I signed up for the study, and why I'm writing this book. We are living through a strange era. Our old certainties are collapsing, and as author Erik Davis has pointed out, the world is getting extremely weird. The fact that psychedelics are going mainstream now is serendipitous, because the very same cognitive, emotional, and spiritual techniques that help us navigate a psychedelic experience can help us navigate these strange times – but we have to use them in the right way.

This book explores how these medicines can change not just our minds, but culture itself. I draw on conversations with rebellious researchers, grounded mystics, and sociologists who study Internet culture; artists, philosophers, indigenous elders, and parapsychologists. I explore what it means for these powerful, transformative medicines to be gaining cultural acceptance at this moment in history. Beyond the hype and the headlines, what does it mean to live in a world in which accessing powerful altered states, mystical visions, and deep experiences of interconnectedness is becoming increasingly mainstream?

Interweaving some of my diary entries from the time, I tie these bigger themes into my own personal journey with the DMT trial. I share personal revelations, paranormal events, encounters with strange entities, and other phenomena that led me to reframe my own conception of reality. I also look at what others experienced and what the researchers discovered during that trial.

In the process, I delve into how the insights, cognitive capacities, and spiritual attitudes that help us navigate altered states may be the same skills we need to transform whole societies. You'll see how these approaches can help us make sense of our own lives, relationships, and work. I also share techniques, meditations, conversational practices, and other methods aside from psychedelics that help us with sense-making.

In Chapter One, I speak to leading scientists and philosophers to explore how psychedelics work, and how they can transform our perception. I investigate the places where science and spirituality overlap in interesting ways, and how this intersection could lead to wider social transformation.

In Chapter Two, I look at the current context of psychedelic use and the reasons we may turn to psychedelics. I map out the 'Big Crisis' we face at this point in history: the complex overlap of the many crises we face as a species, from environmental collapse to political polarization. As well as laying out some of the most relevant aspects of the Big Crisis, I'll share some of the practices and techniques that can help us navigate it more effectively, and explain why psychedelics can provide an essential piece of the puzzle.

In Chapter Three, I inquire into how psychedelics can help us make sense of the other world where we spend much of our waking life: the Internet. I explain how a shamanic perspective on the Internet can change our view of psychedelics, how new religions are forming online, and why psychedelic insights are so useful in changing our collective process of coming to truth online.

In Chapter Four, things get dark. Here, I explore the shadow side of the psychedelic renaissance, and of the incentive structures that govern our economies. I look at the pros and cons of 'psychedelic capitalism,' the negative aspects of mainstreaming psychedelics, and the hope to be found within the murkiness.

In Chapters Five and Six, I take you on a journey into the mystical experiences elicited by psychedelics, which may hold the key to significant and lasting systemic change. I detail how what we're learning from psychedelic science and spirituality can help us play a new social game, weaving together the many conversations that contributed to this book. I also explore some of the practical, small-scale solutions for how we can individually and collectively respond to the challenges of our era.

Something to Remember

One insight that stayed with me most after the trial is that what we are truly missing, and what we need in order to meet the crisis of the times, is a shared reality. Touching this reality is both incredibly simple and profoundly difficult. Simple, because what really binds us together, no matter what our political views or cultural identities may be, is our essential humanity – our hopes, our pain, our longings. Difficult, because our journey back to that place is the greatest journey any of us can undertake. It's full of twists and turns, angels and demons, hidden wisdom and aching deceits.

Psychedelic healing reveals something vital about cultural healing – that we must fearlessly move in two directions at once: deep into ourselves *and* out into the cosmos. Going into ourselves means grappling with the grit of being human – the darkness and blindness, as well as the joy and creativity. That place of humility when we realize we were wrong, and we've hurt someone else. The joy we feel when we truly connect with another person. All the excitement, all the pain, all the uncertainty and longing of being alive and flawed. Death, love, dancing. The whole messy, beautiful drama. It's from that place, and only that place, that we can reach beyond ourselves to the sacred. We have to be rooted in the dirt to reach for the stars.

If we can connect to the fullness of our humanity, we can build a shared reality together that will allow us to meet our shared challenges. But none of us know how to do that yet. And so, our greatest asset is our own curiosity. It is that quality that has led to the psychedelic renaissance, and that led me to the voyage I embarked on during the trial.

It's that same quality I invite you to cultivate as you read this book. Within these pages, you will find ideas and experiences that may feel familiar to you, or weird, or completely out there. I invite you to play with them, to see where they relate to your own reality, and above all, to hold all of them lightly. Because in the places you're about to visit, nothing is quite as it appears. What we know can be revealed as an illusion, and what is left in its place can be more than we ever imagined.

Chapter One

Strange Drugs for Strange Times

An old woman sits in her hut. She eats the medicine with slow, thoughtful bites. As she waits for the spirit to visit her, she thinks of the changes she's seen. The dry earth, the thirsty trees. Her tribe will need to move soon, but no one knows where. They look to her to tell them, and she looks beyond herself.

Soon, she feels their presence – spirits crowding the hut, swimming in and out of her. She sheds her body like a skin and travels with them, far up into the sky, where she can see how things weave together. Where the rains will come, where the antelope will run. They tell her other things, too. Things that she can never put into words. Things that can only be danced or howled. When they're gone, she returns to tell the tribe.

The woman I have just described does not exist. And yet, she does. She is the archetype of the shaman. I present her here not to comment on a particular group or practice, but to represent an amalgamation of many traditions around the world.

Derived from a Manchu-Tungus word (from the region of Eastern Manchuria and Siberia, where Tungusic languages are spoken) meaning 'they who know,' the shaman is part doctor, part strategist, and part visionary. Shamans play an essential role in many smaller societies: that of the person who mediates between the world we can see, and the world we can't. They navigate the social dynamics of the tribe, and the larger dynamics of a vast ecosystem of intelligences belonging to the natural world.

Shamans are often confronted with spirits who show up as tricksters and shapeshifters. And so, perhaps it's no surprise that even the definition is contentious, and what the word 'shaman' should mean, and to whom it should apply, is always shifting. What many anthropologists agree on is that shamanism is the oldest form of religious practice we're aware of. It involves the use of practices to alter our day-to-day consciousness and experience new ways of perceiving. Usually, this is done by communicating with other intelligences, whether the spirits of nature, ancestors, or other, stranger, entities.

A Shipibo-Conibo *vegetalista* is a type of shaman in the Peruvian Amazon who uses their knowledge of plants to effect change; they may use ayahuasca in ceremonial space to diagnose and heal illness. A Tungus shaman uses the distinctive red-and-white polka-dot mushroom known as the fly agaric (or by its proper name, *Amanita muscaria*) to travel beyond their own bodies. Far from the snows, the San people of Botswana enter an altered state through trance dancing, a ceremony that is recorded in ancient cave paintings and still takes place today. Thousands of miles north, the Punu and Mitsogo peoples of Gabon use the psychedelic medicine iboga as part of a spiritual discipline known as Bwiti. Around the world, across all times and places, people have used practices like drug-induced states, rhythmic drumming, fasting, isolation, and breathing to reach beyond themselves.

The modern psychedelic renaissance can feel far removed from these ancient practices, but it isn't. Anthropologist Bia Labate, who heads up a nonprofit psychedelic charity called the Chacruna Institute, has spent her career focused on finding equitable solutions bridging the gap between indigenous wisdom and the modern psychedelic movement in the West. She explained to me that 'there is a deep continuity between shamanism, the underground ritual, traditional practices, underground therapeutic circles, and the above-ground clinical-trial FDA approach. All of this is part of the same movement, either because we're using the same substances... or because we studied the practices [from indigenous groups] and got the idea that these substances heal.'[1]

Labate gave an example of how in the 1950s, researchers in Canada participated in Native American teepee ceremonies and observed how indigenous people used peyote to treat alcoholism. This inspired the researchers to propose a similar study with LSD in the 1960s. The West first became aware of the sacramental use of magic mushrooms thanks to a man named R. Gordon Wasson. An investment banker and amateur mycologist, Wasson in turn discovered them through María Sabina, a Mexican Mazatec shaman. Had it not been for Sabina, psychologists in the 1950s and 1960s might never have started experimenting with these medicines.

During that era, dozens of studies started to reveal just how powerful psychedelics could be as clinical and psychiatric medicines. Initially, many called them 'psychotomimetics,' as they were believed to mimic psychosis. The term 'psychedelic' was created by Aldous Huxley and Humphry Osmond to capture what they believed these substances really do: show us our own minds – and more than that, what lies beyond our own minds.

In the 1950s, many clinicians hailed LSD and other psychedelics as a revolution in psychiatry. Psychologists began experimenting with LSD to treat alcoholism, depression, and other conditions. However, these powerful molecules have a habit of taking us where we aren't expecting to go. And in the 1960s, psychedelics left the lab and entered public awareness, making a lasting impact on social values, music, spirituality, and pop culture in the West.

While many have attributed aspects of the art that drove the late 1960s and early 70s to psychedelics, writer and historian of psychedelic culture Erik Davis believes that the most lasting effect of the 'psychedelic 60s' was spiritual.[2]

Just as psychedelics can help individuals encounter the complexity of light and darkness within themselves, their explosion into the counterculture in the late 1960s brought both benefits and harms. Psychedelics opened a generation up to different forms of spirituality, from yoga and mindfulness to group work and new forms of therapy. However, the West's first encounter with psychedelics was no utopian picnic. Many suffered psychological damage from unsafe use, and the famous Summer of Love of 1969, centered in San Francisco's Haight-Ashbury district, soon gave way to heroin use and broken lives.

One of the key figures during this time was Timothy Leary, a Harvard psychologist who became an advocate for psychedelics as cultural deconditioning agents. He famously asked young

Americans to 'turn on, tune in, and drop out.' This idea of rejecting your culture and recognizing it as something constructed by someone else was another powerful impact of this early counterculture, interwoven with changing ideas about power, boundaries, and sexuality. Psychedelics became associated with the anti-war and civil-rights movements. This development was so challenging to the status quo that Richard Nixon called Timothy Leary 'the most dangerous man in America.' In 1971, Nixon's government banned cannabis and psychedelics. As Nixon's domestic policy chief John Ehrlichman would recount years later:

"The Nixon White House... had two enemies: the antiwar left and black people.... We knew we couldn't make it illegal to be either against the war or black, but by getting the public to associate the hippies with marijuana and blacks with heroin, and then criminalizing both heavily, we could disrupt those communities. We could arrest their leaders, raid their homes, break up their meetings, and vilify them night after night on the evening news. Did we know we were lying about the drugs? Of course we did."[3]

Medical trials examining the efficacy of psychedelics in treating depression, anxiety, addiction, and other indications ceased almost overnight. For the next 30 years, there was a research blackout, but the psychedelic flame kept burning. Pioneering researchers like Amanda Feilding and Rick Doblin kept research going against significant odds. Psychedelic philosophers like Terence McKenna inspired people to continue exploring their consciousness in the underground, but for mainstream society, psychedelics were taboo.

A Colorful History

The relegation of psychedelics to underground circles began to shift as the mental-health crisis worsened. Starting in the 1990s and early 2000s, scientists began to revisit the research from the 1950s and 60s to propose new experiments, often aimed at determining whether psychedelics could in fact become new psychiatric medicines.

One of the most influential early studies came in 2006, when a team at Johns Hopkins led by Roland Griffiths published an influential paper titled 'Psilocybin can occasion mystical-type experiences having substantial and sustained personal meaning and spiritual significance.' In it, the team reported that an astonishing 67 percent of participants rated their experience with psilocybin to be either the most meaningful experience of their lives, or in the top five, alongside the birth of a child or the death of a parent.[4]

This capacity for psychedelics to elicit profound experiences of connection with reality is now seen by many clinicians to be one of the most important ways they heal. Griffiths was building on another famous study that happened at Harvard University on Good Friday of 1962. In the basement of a divinity school, researcher Walter Pahnke gave capsules to 20 Protestant divinity students; the capsules contained either psilocybin or a placebo. Those who received psilocybin reported profound mystical experiences that changed their lives.

In a study published in 1991, Rick Doblin interviewed the participants almost 30 years later. Overwhelmingly, they reported that the experience had had lasting and sustained changes, helping them to 'resolve career decisions, recognize the arbitrariness of ego boundaries, increase their depth of faith, increase their appreciation of eternal life, deepen their sense of the meaning of Christ, and heighten their sense of joy and beauty.'[5]

The connection between psychedelics and profound mystical experiences isn't a discovery. It's a rediscovery.

As well as having roots in shamanism, mystical experiences caused by psychedelics might have played a role in the development of Western culture. Perhaps the most famous example is the Eleusinian Mysteries of ancient Greece, which took place every year for almost 2,000 years. Initiates participated in a mysterious ritual which was so profound that none were allowed to speak of it. It involved participating in the myth of the abduction of Persephone from her mother Demeter by Hades, which in turn represents the change of seasons and is linked to the agricultural cycles that allowed civilization to keep going.

The mysteries were incredibly influential on thousands of people, including Sophocles, Aristotle, Plato, and Cicero. 'Nothing is higher than these mysteries,' Cicero wrote. 'They have not only shown us how to live joyfully but they have taught us how to die with a better hope.'[6] As Brian Muraresku points out in *The Immortality Key*, there is strong evidence to suggest that the *kykeon* that participants drank during the ritual contained a psychedelic compound. He argues that psychedelics might have played a key role not just in the mysteries, but in the development of early Christianity, as well.[7]

This deep interconnection between psychedelics and the culture and practices around them makes sense when we examine the latest research. Psychedelics are complex both psychologically and neurologically, but a leading theory is that they act as connectors. That is, they connect us to ourselves and to the world around us. And when we take a psychedelic, different parts of the brain that normally don't interact with each other are brought into contact. Psychologically, we connect to memories, feelings, and behaviors we aren't normally aware of. Spiritually, they connect us to the profound mystery of being.

Author Richard Doyle describes psychedelics as 'ecodelics.' In his book *Darwin's Pharmacy: Sex, Plants, and the Evolution of the Noösphere*, he describes how after reviewing thousands of trip reports, he noticed that the most common theme was an 'ecodelic' insight, which he describes as 'the sudden and absolute conviction that the psychonaut is involved in a densely interconnected ecosystem...'[8] This can be our immediate environment, the city we live in, the forest around us, the entire ecosystem of the planet, or all of them simultaneously.

Psychedelics also connect us to one another – to the stories we share and the myths we live by. To understand why this makes them so promising when it comes to helping us make sense of the times we live in, we need to look at how they achieve this.

How Psychedelics Change Our Perception

As we've seen, humans have been using psychedelics for thousands of years to expand our perception of reality. Shamanism, including the use of medicinal plants to attain wisdom, is widely regarded as one of the first forms of human spirituality and healing. Cognitive scientist John Vervaeke has pointed out that shamanism is often associated with a phenomenon called 'soul flight.'[9] The shaman might have the experience of transforming into a bird, or leaving her body and seeing the world from above. In doing so, she can identify previously hidden patterns, such as where to find food, when it's going to rain, or a new insight into mediating a dispute in the tribe. It's no coincidence we still talk about 'getting high.'

Vervaeke sees this as a cognitive process of 'zooming out' from our normal frame of reference. He uses the example of taking off a pair of broken glasses. Normally we see *through* our frames, but when we zoom out we're looking *to* our frames. Where are they cracked and smudged? Where are they warping how we perceive? We can then put our glasses back on and 'zoom in' through a new set of perceptual frames and see the world differently.

This process may be vital for making sense and finding meaning. Vervaeke's research is centered around a concept called 'relevance realization.' The idea is that human intelligence is based on our ability to determine what is *relevant* to us from all the things around us that are *salient*, or calling for our attention. As you're reading these words, there is a lot of salient information around you that you're tuning out. These may include slight fluctuations in temperature, sounds or smells in your environment, and the constant stream of your own thoughts and emotional experience.

From all of this salient information, you are able to decide what's relevant to you and your goals, and to focus on that. However, this very same ability is also what gets us stuck. We can make the wrong things relevant and get lost in our own misconceptions, delusions, and narrow frames of reference. People often accuse one another of being biased, but intelligence *is* bias; it's a process of honing your attention so that you don't get lost in a sea of information. Sometimes we use this bias well, and sometimes we don't. No matter how smart you are, we're all in the same boat, because the same cognitive machinery that makes you smart also makes you stupid. To become less stupid, and open up to new ways of seeing and being that can help us make wiser choices, we have to make and break our frames regularly. This is what enables us to have those 'aha!' moments in which we realize what we weren't perceiving, what was *outside our frame*.

We can see why this matters by considering what happens when our frames become too narrow. Neuroscientist Marc Lewis has argued that addictions might be an example of what he calls 'reciprocal narrowing.'[10] What that means is that our frame of reference – what we can see as a possibility for ourselves, our own lives, and our behavior – gets narrower and narrower until we're stuck in a kind of perceptual box we can't get out of. I would argue that ideologies function in a similar way: Any all-encompassing narrative narrows our ability to think flexibly until, eventually, we can only see the world through a narrow frame. Narrow frames always give us an incomplete view of one another and reality, and they're rarely helpful.

Vervaeke has suggested that if there's such a thing as 'reciprocal narrowing,' there is most likely also an experience of 'reciprocal opening.'[11] This is an experience of expansion, in which our frames of possibility zoom out to get wider and wider. This can take place in many different ways: dancing, meditating, psychedelics, or even a wonderful and nourishing conversation. However, we can't expand our view on the world indefinitely. Vervaeke has theorized that our brains reward us most when we have an 'optimal grip' on reality; when we're neither too zoomed out or too zoomed in, but just right.

This may be why the mystical experience psychedelic researchers are often hoping to elicit in study participants isn't a magic bullet. We can experience a deep unity with nature, or a sense of being tiny yet significant in a vast cosmos, or a dissolution of our fixed identity and a merging with everything. But then, we have to come back and apply this new frame to our lives. We have to do the hard work of changing how we think on a daily basis and making different choices. We need others around us with whom we can connect and make sense of the experience together. Psychedelic researcher Betty Eisner called this our 'Matrix' and argued that it should be added to the famous set, setting, and dose as a core aspect of healthy and sustainable psychedelic healing.[12]

Sometimes, our environment doesn't allow for this. It's hard to reconcile a feeling of unity with the universe while you're filling out a spreadsheet at work. And even though psychedelics may reveal to us the changes we need to make, many of us can't always afford to do the things we need to feel better. Not only do we need the right kinds of integration techniques – we need the right kind of society to support integration of these powerful experiences.

Many spiritual traditions that use practices to expand our frames have a built-in sense of community and meaningful action in the world for this very reason. The Buddhist practice of mindfulness is a process of looking to your senses instead of through them, and thereby paying attention to the way you're seeing the world, or the frames you've put around it. In many Buddhist traditions, people have this experience within a *sangha* – a community of others walking a similar path. Our community can help us to 'zoom in' to our new perspectives and make sure we don't get stuck in one of the many cognitive loops humans are susceptible to. Likewise, core to Buddhism is bringing the *dharma*, the teachings of the Buddha, into how we live our lives collectively.

Right now, our scientific-materialist culture is approaching psychedelics in the same way we approach everything else: with an atomized, disconnected approach that separates the 'sick' person from their context and sees them not as a part of a complex social and natural ecosystem, but as an object to be fixed. However, as I'll explore later in the book, many clinicians and retreat centers are also working on ways to solve this – to create a more holistic approach that allows us to make significant and lasting changes that don't just help the individual, but also serve the community. When we do that well, we begin to cultivate wisdom: the process of learning how to effectively apply our knowledge.

The idea of a truly holistic approach to psychedelic healing means holding multiple perspectives at once, recognizing that both neuroscience and spirituality can help us make sense of the experience in different ways. My thinking in this area has been greatly influenced by Integral Studies, the philosophy created by Ken Wilber. Integral helps us to hold multiple perspectives at once, and understand that they can coexist without one being dominant. Scientific inquiry is focused on knowledge that we can agree about together: facts and propositions. However, it is not set up to understand or comment on what it's like to be you.

Similarly, my subjective experience can't be used to comment on the reality we all share.

With this in mind, for a more comprehensive picture, we need to hold science and spirituality together, linking our subjective experience with neuroscience and vice versa. When we do that, we can start to marry consciousness and neuroscience in new ways.

What we're learning from psychedelic neuroscience with respect to the connection between the brain and our capacity to adopt new perspectives is fascinating. The foremost psychedelic neuroscientist in the world is Robin Carhart-Harris. His research supports the idea of psychedelics as 'reframing agents.' He and others have suggested that psychedelics lead to something called *neuroplasticity*: our brain's capacity to change and rewire itself.[13] When we spoke, he explained that one reason psychedelics may be so effective in treating mental illnesses is that they break us out of fixed, rigid ways of seeing ourselves and the world.

Carhart-Harris has suggested that symptoms like depression or anxiety could have the common cause of 'reinforced, maladaptive habits or biases.'[14] Research suggests that psychedelics work in part by catalyzing a window of plasticity in the brain. Plasticity is the brain's ability to change itself; in this process, neurons rewire to form new connections and release old ones. Psychedelics may create an opening for the creation of new neuronal links, frames, connections, and behaviors that can get us out of these reinforced habits. Psychedelics aren't alone in doing this. Carhart-Harris's and others' research suggests that pressure on an organism might also elicit neuroplasticity. This may be the neuroscientific correlate of a piece of wisdom I once received from Zen Master Doshin Roshi, who asked, 'When do we change? When we have to.'

However, it's not psychedelics alone that can change our perception; it's the drug in combination with a practice like therapy. The positive clinical-trial data driving the psychedelic renaissance comes from a union between pharmacology and a practice of deep therapeutic holding.

Psychedelic therapist and author Maria Papaspyrou has been at the forefront of figuring out how to combine therapy with the psychedelic experience. When we spoke for this book, she explained just how delicate and complex psychedelic therapy is as an approach, and how different to other forms of therapy we might be used to. 'The therapeutic process involves a very deep surrender for both the therapist and client,' she explained. 'Stanislav Grof spoke about the 'healing intelligence,' which is the deeper layers of knowing that the participant will open up to. The therapist is not there to influence or guide that, but to move with it.'

Once we've had this transformative experience, we need to integrate it. Integration, Papaspyrou told me, 'is about meaning-making: making sense of the experience and finding a way to make some kind of use of it in your everyday life.'[15] This brings us back to Vervaeke's glasses metaphor: Integration is like putting your new glasses on and making sure that your reframing was effective and beneficial to your life.

Sometimes, the insights we have during a psychedelic experience help us reframe our own lives. We may see that we're not dealing with anger well, or we aren't expressing ourselves authentically at work. We may realize we need to exercise more, or bring more creativity into our lives. Other times, we have an experience that completely reframes what we thought reality was. DMT, perhaps more than any other psychedelic, can lead to this radical shift in perspective.

What Is DMT?

DMT is a very unusual molecule. It's found in a number of plants and mammals, including human beings and orange trees. The famed ethnobotanist Dennis McKenna has argued that it may be in every plant and every animal; we just haven't looked.[16] Your body is full of DMT. It might be released during birth and death, and may also play a role in dreaming. It's also a Schedule I substance, so don't mention any of this next time you're going through customs.

So, why, if this molecule is so widespread, don't you go into an altered state every time you drink orange juice? Firstly, it's because DMT is present in very low quantities. But it's also because there are enzymes in your gut known as monoamine oxidase (MAO) that render it orally inactive. DMT is one of the active ingredients in ayahuasca, which is a combination of a vine (*Banisteriopsis caapi*), a DMT-rich leaf (*Psychotria viridis*), and often, a variety of other medicinal plants. The caapi vine is rich in two important molecules, harmine and harmaline. These are psychoactive by themselves and also act as powerful inhibitors of the monoaminoxidase enzymes that disable DMT in your gut and allow it to be orally active.

Ayahuasca has been used by people around the world for everything from gaining spiritual insight to treating alcoholism and opioid addiction. In the last 10 years, it's become wildly popular with people in the developed world. Celebrities like Will Smith talked about the profoundly healing effect it had just months before he slapped Chris Rock at the Oscars, then later shared how he'd had a vision on ayahuasca that he would do something to ruin his career. This is a good metaphor for the confusion that has happened as ayahuasca has become mainstream. Tourists regularly flock to Peru, Colombia, and Ecuador to participate in ceremonies – and as we'll see later in the book, this explosion of popular interest is a double-edged sword, as we can't always integrate what we experience into the world we live in.

Like the ayahuasca brew that contains it, DMT also has pop-culture notoriety. It has come to represent something deeply mysterious and profound. For thousands of years, indigenous cultures (and countercultural explorers in the West) have found ways to ingest DMT. Archaeological evidence suggests that a DMT-containing snuff, likely from the plant *Anadenanthera*, has been used in South America for more than 4,000 years.[17] It was traditionally blown into the nostrils, where the DMT isn't broken down by your MAO enzymes.

After psychedelics hit public awareness in the 1950s, people in the West began to extract DMT from plants and vaporize it for a similar effect. As with the snuff, this method of ingestion brings on the experience very rapidly – and many report it as one of the most profound, shocking, and inexplicable experiences of their lives. It is short-lived, lasting about 10 minutes, but very powerful.

As soon as the vapor hits your lungs, the world changes. A sharp ringing fills your ears. You may have the sensation of leaving your body. Intricate geometric patterns explode around you. Spirits and entities of all shapes and sizes emerge from nowhere – some curious, some loving, some cold and calculating. And some people experience the presence of a 'teaching intelligence' that provides insight and even knowledge.

More than any other aspect of the experience, it is the encounter with entities that has captured people's imaginations. How these entities are perceived differs from culture to culture – from insectoid aliens and cartoonish ghouls to faeries, spirits, and jesters. Ask people from different cultures what they are, and you'll get a lot of different answers. To a Jungian psychologist in New York City, they may be representations of our unconscious subpersonalities; to a neuroscientist in Tokyo, simply the result of a confused brain; to a shaman in Peru, the spirits of the dead. Others theorize that they're alien intelligences in another dimension that exist independently of us.

Regardless of the nature of these entities, human beings have been communicating with them for tens of thousands of years. Some scientists believe DMT may be naturally released in large quantities at death and theorize that it is involved in the mystical near-death experience (NDE) consistently reported by people across time periods and cultures. In many NDEs, people perceive a bright light and reconnect with loved ones. Others have argued that the spontaneous religious revelations of Biblical prophets may have been influenced by a natural release of DMT by their own brains. In his book *Supernatural*, Graham Hancock argues that faerie abductions, UFO encounters, and DMT experiences may very well be the same phenomenon, and suggests that these entity experiences may even be the result of a release of endogenous (meaning 'in the body') DMT.[18] However, scientists debate if the brain can release the amount of DMT required for this kind of experience.[19,20]

Setting aside neurobiology, reading the reports of people who've taken DMT shows why this link between the entities of folklore and UFO encounters has been made by so many. The first clinical trial investigating DMT to take place after research was banned in the 1970s was run by Rick Strassman in the early 90s. In fact, this was in many ways the first study of the current renaissance in psychedelic research. Strassman would go on to write *DMT: The Spirit Molecule*, documenting both his quest to receive FDA approval for the trial, and what happened during it. The book and a subsequent documentary have become legendary in the psychedelic world, due in part to the extraordinary experiences people reported.

A 45-year-old blacksmith named Karl described what he encountered during one of his dosings: 'There were a lot of elves. They were prankish, ornery, maybe four of them appeared at the side of a stretch of interstate highway I travel regularly. They commanded the scene, it was their terrain! They were about my height. They held up placards, showing me these incredibly beautiful, complex, swirling geometric scenes in them.'[21]

Another participant described something more sci-fi in feel: 'There was a space station below me and to my right. There were at least two presences, one on either side of me, guiding me to a platform. I was also aware of many entities inside the space station – automatons, android-like creatures that looked like a cross between crash dummies and the Empire troops from

Star Wars, except that they were living beings, not robots. They seemed to have checkerboard patterns on parts of their bodies, especially their upper arms. They were doing some kind of routine technological work and paid no attention to me.'[22]

Participants regularly encountered entities that seemed actively interested in humanity. Jeremiah was 50 at the time of Strassman's study and was training as a counselor, having just left a career in the armed forces. He described entering a high-tech nursery during one of his dosings, and encountering an entity that regarded him with 'a sort of a detached concern, like a parent would feel looking into a playpen at his one-year-old lying there. As I went into it, I heard a sound: hmmm. Then I heard two to three male voices talking. I heard one of them say, "He's arrived." I felt evolution occurring. These intelligences are looking over us. There is hope beyond the mess we are making for ourselves.'[23]

What is most striking about the DMT experience is that people don't often interpret these experiences as drug-related, but as encounters with a cohesive, independent reality. Jeremiah told Strassman that 'DMT has shown me the reality that there is infinite variation on reality. There is the real possibility of adjacent dimensions... it's not like some kind of drug. It's more like an experience of a new technology than a drug.... It's not a hallucination, but an observation. When I'm there, I'm not intoxicated. I'm lucid and sober.'[24]

In an academic paper published in 2020 by Alan Davis et al, the researchers examined the responses of around 2,500 people who had filled out a questionnaire about their encounter with DMT entities. When asked about the attributes of the entity,

96 percent of respondents reported that they believed it was both conscious and intelligent. 78 percent reported that it was benevolent, 70 percent saw it as sacred, and just over half (54 percent) experienced it as having 'agency in the world.' While most reported the entities as either positive or neutral in the way they perceived the subject, a smaller amount reported that they were 'negatively judgmental' (16 percent), or even malicious (11 percent).[25]

The experience also had an effect on people's spiritual outlook. The study reports that 'approximately one-third (36 percent) of respondents reported that before the encounter, their belief system included a belief in ultimate reality, higher power, God, or universal divinity, but a significantly larger percentage of respondents (58 percent) reported this belief system after the encounter.' A 2018 study by Chris Timmermann argued that the DMT experience closely corresponds to the near-death experience, and a 2021 field study by Pascal Michael and David Luke supports this to an extent, though the authors note that people reported experiences that more often resembled UFO-abduction reports.

A larger study published in 2022 by David Lawrence et al looked at almost 4,000 reports of people in a DMT community on Reddit (r/DMT) taken over a 10-year period, in order to examine whether these experiences may have therapeutic application. They yielded a similar amount of shared entity experiences as the Davis study, with around half the reports describing entities of varying qualities and intentions – the most common being an archetypal female entity or goddess (24 percent of reports) and one of the least common being an entity wearing a tuxedo and top hat, which 11 people reported (0.6 percent).[26]

While this research is important in gaining a more objective view of the experience, something I often contemplated during the trial was just how ill-equipped neuroscience is to make sense of the subjective experiences people are having on DMT.

At the same time, as an individual going into a psychedelic experience, you can practice a kind of internal science. We can bring a similar curiosity, discernment, and rigor to exploring our inner worlds as we can to the world we see around us. Navigating the DMT experience safely is a skill that requires preparation, cognitive training, and experience.

Preparing for the Psychedelic Journey

This is why all of the participants in the DMT continuous-infusion trial I was on were experienced with psychedelics. It was a dosing trial to figure out what the dose should be for a wider study, so we were the guinea pigs for the guinea pigs. Eleven of us began together, and almost a third would ultimately drop out, in large part because of the intensity.

A prerequisite for the study was to stop taking any recreational drugs, and I decided to stop drinking alcohol, as well. In order to map my process and my preparation, and to integrate my experiences, I took detailed diaries throughout the trial. These weren't just of my dosings, but also of events that took place between dosings; these events were often stranger than the DMT experiences themselves.

I also had biweekly sessions with a facilitator and coach named Trish Blain so that I could both prepare for dosings and get help making sense of what I experienced. Blain helps people navigate non-ordinary states of consciousness with a model she's developed called The Four Forces. It divides human experience into four basic desires (expression, growth, purpose, and connection) and uses them as a map to untangle what we're experiencing in different spiritual practices and non-ordinary states, as well as day to day.

The 'kenshō' experience of Zen Buddhism, of becoming one with all things, can be seen as the ultimate moment of *connection* with a capital C. No longer are we separate individuals – rather, we dissolve into the vast nothingness of the universe. But we can also have a very different mystical experience in which our single point of consciousness becomes the whole universe, which is the ultimate in *expression*.

This is an experience I've had meditating on LSD – an experience of 'all is one and you are it,' in which my unique expression of consciousness was one with the universe. Alternatively, we might have a peak experience in which we see the world as full of meaning, intention, and direction. Everything is revealed, we see underlying patterns behind things, and know that everything is coming together for a higher *purpose*. Or, we might have an ecstatic experience in which we're filled with life-force energy, or *growth*. Tantric experiences, like a kundalini awakening, are ones of tapping into a wellspring of aliveness and expansion.[27]

As well as preparing for different types of mystical experience, I drew on my own training to incorporate practices I hypothesized would help me travel as deeply as possible, including mindfulness and inquiry. Inquiry is a specific kind of talking meditation, a practice I learned and fell in love with in a counseling training I received from therapists Rafia Morgan and Turiya Hanover.

As Morgan explains it, implicit in the concept of inquiry is an attitude of 'I don't know.' In many ways, it's the same process the scientists on the trial were on, except in this case, it's applied to unraveling the threads of our inner worlds. It would prove indispensable in navigating my dosings. Inquiry as I learned it comes from a tradition called the Diamond Approach, or Ridhwan, a marriage between depth psychology, Sufism, and other spiritual traditions. As one of its founders, A.H. Almaas, explains, 'To contact the deeper truth of who we are, we must engage in some activity or practice that questions what we assume to be true about ourselves.'[28]

Inquiry helps us to do this, and it comes in many forms. Often, two people sit opposite one another while one inquires into the nature of their experience in that moment, while the other listens without interrupting. Above all, inquiry teaches us how to follow our curiosity.

We speak, and as we do this, we notice what's happening. As I write that sentence, I notice a pressure to describe the process. And having expressed that, I feel a tightness in my chest. I might take a moment to be with that tightness and see where it leads me – perhaps to an image or sensation, or a subtle emotion. When we inquire, we aren't telling a story. Telling a story, we're always trying to get to the climax. In inquiry, the process *is* the climax. It's a transformative practice of unfolding that helps us unravel the truth of our lived experience. Practiced enough, inquiry becomes not just a technique, but an attitude to life.

I also drew on an overall attitude toward spiritual experience and personal growth I learned through Path of Love, a seven-day process run by Morgan and Hanover. The process involves group inquiry, shadow work, exposure therapy, and ritual. Participating in and facilitating this kind of work, I came to the perspective that developing a healthy ego is just as important as growing beyond it. For me, spirituality is a process of developing authentic human connection, as well as connecting to what lies beyond us. Such a process requires grit, self-acceptance, and aching intimacy. It requires a willingness to be as authentically, messily human as possible – as well as accepting of our limitations when we can't. Spirituality offers a way of being in the world that's as eager to be down in the muck as connected to the stars.

Throughout the trial, I would refer back to these techniques and attitudes as I kept a diary to make sense of what I was experiencing. At points throughout this book, I have included sections of my dosings drawn from these diary entries. At times, I wondered whether to do so, as some were very personal, and I am acutely aware that I don't want to be perceived as making any reality claims based on my own subjective experiences.

But the power of the psychedelic experiences lie in their subjective effects, and to understand how they can help us make sense of the world, we have to look at what actually happens during them. Clinical-trial papers are important, but they don't tell us much about the experience itself. They don't explain *how* psychedelics heal, or the process whereby we change our perspective. They can't capture the quality of that experience: the astonishing, loving, and powerful way they connect us to our own minds and souls.

Realer than Real

There is a powerful tension between the subjective experiences we have on psychedelics, and the objective data that scientists are looking to gather by observing those experiences. From the lens of materialist science, your consciousness is a byproduct of what's happening in your brain. In a sense, from this worldview you aren't really anything more than meat that's temporarily aware of itself.

By extension, everything taking place in a DMT experience is just your brain on drugs. But that is not what most people come back believing after their DMT experiences. In fact, psychedelics seem to shift people away from this perspective and toward one that views consciousness as a fundamental quality of the universe. And yet, to remain objective, the scientific study of DMT has to twist and turn to capture the intensity, weirdness, and commonality of these experiences into a frame too small to hold them.

When I spoke to Rick Strassman about this, he told me that it isn't just our scientific tools that make it hard to make sense of what's going on, but our philosophical expectations, as well. While many of his participants had mystical experiences, they didn't get the *type* of mystical experience they, or he, were expecting. Strassman and many of the participants were Zen Buddhist. He explained that they were expecting a kind of *samadhi* experience. Samadhi is a state described by ancient Hindu yogis in which union with the divine is achieved, and the individual self dissolves into the pure emptiness underlying reality.

But people didn't get emptiness. They got a whole lot of *something*. Many experienced a different, no less powerful, kind of mystical experience that took them beyond their own individuality, but was fundamentally *relational*. It didn't involve a dissolution of individual identity, but was a process in which people related, communicated, and interacted with a new reality full of other intelligences in an experience that many reported felt 'realer than real.'

This relational experience was so consistent in the dosings that Strassman would later search for new spiritual frameworks to try to make sense of it. What he found was that the tradition of prophecy in the Abrahamic religions provided the best he could find. Strassman defines prophecy loosely, explaining that any visionary experience recounted in the Bible could fit the bill – a concept he explores in detail in his book *DMT and the Soul of Prophecy: A New Science of Spiritual Revelation in the Hebrew Bible*. As he describes it, the experience is closer to Moses being instructed by the Burning Bush than it is to a Zen monk meditating on a mountain.[29]

The psychedelic encounters that people reported were relational in that they experienced a back-and-forth 'dialogue' with what they were encountering. This aspect of the DMT experience was something I was acutely aware of in the lead-up to the trial, and the reason I believed DMT was the psychedelic best suited to making sense of the chaos of the times. It has the potential to bring us into contact not just with new ideas, but intelligent entities and entire civilizations that may be able to give us a new perspective on our own species. In such an experience, we maintain our agency (our ability to act) while actively exploring a radically different landscape of possibility.

The question of whether the entities people encounter on DMT are 'real' is a fascinating one, with significant philosophical implications. This is known as their ontological status. Ontology is the philosophy of what exists, and the nature of being. Strassman argued that the 'realness' of entities is inherently unknowable. Likewise, David Luke, a parapsychologist and one of the world's foremost DMT researchers, made a similar point when we discussed the matter. He told me that even if you were to, for example, go into a DMT experience with a complicated math equation you didn't know the answer to (which some have tried) and try to get help solving it, you wouldn't prove anything. Even if you came out with an answer, there are too many variables. Perhaps you received it from some part of your own mind. Perhaps a spirit told you. Perhaps it was telepathically sent to you from a mathematician across the globe.

I will return to the debate around what exactly the entities are later in the book. Going into the dosing, I was trying to hold an attitude of inquiry around them. I was also feeling eager and restless. My first dosing two weeks earlier had turned out to be a placebo, which was both disappointing and very obvious. It also reminded me of the awkward fit between psychedelic research and the double-blind placebo controlled-trial model, usually the gold standard in research. For example, in a 2022 study that showed the effectiveness of psilocybin in treating alcoholism, researchers revealed that 95 percent of participants were able to identify whether they got the placebo or psilocybin.[30]

Having had my own placebo experience, I was already familiar with the dosing room when I arrived. I walked into a simple hospital room that the team at Imperial had done their best to soften. It was faintly lit by a salt lamp, with a red cloth draped over one wall. Opposite me was a large, dentist-style chair. I sat down at a desk with Lisa Luan, the PhD student responsible for the day-to-day running of the trial. She would sit beside me during every dosing and act as my main point of contact throughout the process. She explained that after a series of neurological tests, I would lie back on that chair and a medic would insert two needles into my arms, held in by cannulas. One needle would connect to the DMT pump, while the other would take blood at regular intervals.

As Lisa installed the 200 suckers of the EEG cap onto my scalp (which I started calling the Squid Helmet), Chris Timmermann conducted a pre-dosing interview. Chris is one of the foremost neuroscientists investigating DMT, and he was the lead researcher on the trial. When he asked me why I'd signed up for the study, I told him I wanted to try and come back from the DMT experience with useful insights about the state of the world; I also admitted to him my desire to communicate with entities.

I also told him that I was hoping for a metaphysical experience instead of something personal. My early psychedelic experiences used to be metaphysical, full of fascinating insights, but ever since I'd started doing inner-development work and therapy, they had become more psychological, which is a good example of just how influenced our experiences are by our expectations and mental frameworks. In my case, I was looking to shift gears into a different frame. I was tired of the endless work on 'project me' that accompanies modern therapy and personal growth. I was looking for more elves, less introspection.

After the interview and the questionnaires, I lay back in the chair wearing the EEG helmet with two cannulas in my arms. Surrounded by scientists with spacey music playing from the speakers, the whole thing felt very sci-fi. I was going boldly where no one had gone before.

Lisa helped me put the blindfold on over the EEG helmet. She set a pair of headphones over my ears. Darkness. Spacious music filled my head. A recording of Chris's voice asked me to give intensity and anxiety ratings on a scale of 0 to 10, with 10 being the most intense. I was reminded that during the experience I would need to reply out loud. If I missed three in a row, they would stop the infusion. If it got too intense, I could ask to stop at any time. They covered my eyes and ears. I waited for the dose.

Information from Beyond

If the idea that we can receive actionable information from seemingly autonomous entities in a drug-induced state seems outlandish, that's because it is. However, it may also be deeply rooted in the deepest parts of how we think and know as human beings. We can see why when we set aside the unanswerable question of the reality status of entities and look to the cutting edge of cognitive science.

A simple definition of cognitive science is that it's the study of how we think and know. Traditionally, scientists saw this as a function of the brain, and because we've been influenced by 400 years of mind–body duality, our society is still dominated by this perspective. In the West in particular, we have a culturally conditioned sense that we are disembodied intelligences, with our primary awareness existing somewhere in the forehead. This is a complete fantasy. The latest cognitive science is known as 4E Cognitive Science. It argues that, far from being located solely in the brain, our cognition is *embodied, embedded, enacted,* and *extended*.

It is *embodied* because it is inseparable from our experience as a physical body. Researcher Barbara Tversky has shown how the way we physically move through space changes how we think and how we solve problems.[31] Our concepts and ideas are inseparable from our gestures, feelings, and sensations. We can see hints of this in our own language when we talk about 'moving on' from a difficult period, or 'taking a step back' to reevaluate a situation. When we admire someone, we 'look up to them,' and when we're feeling excited about an idea, we talk about 'leaning into it.'

Our cognition is also *embedded* in the world, inseparable from the environment to which we've adapted. A whale is embedded in the ocean, and it doesn't make sense to conceptualize it as an animal outside of that environment. Cognition is also *enacted*: Everything you do impacts the world around you and creates new realities that in turn constrain or open up possibilities for you. If you save money, you are opening up possibilities that wouldn't have been there if you hadn't.

Finally our cognition is *extended*; it moves through and responds to other people, technology, and the environment. Studies have shown that, as human beings, we perceive our tools as an extension of our bodies. For example, we don't have to think about how wide our car is getting through a narrow street; we just know. In a sense, the car and our body meld together.[32,33]

In *Cognition in the Wild*, anthropologist and sailor Ed Hutchins explores the question of who navigates a ship. It isn't a single person, he argues, but a whole network of people and their technologies. You could ask the same question about who built your phone or your e-reader. They were created by a 'distributed cognition' of many minds and hearts and hands – an intelligence that extends across countries and interacts with technology to create. All of our big problems are solved by this kind of distributed cognition.

In case four E's aren't enough, John Vervaeke has argued for two more. The first is *emotion*. In a course I ran with him titled 'Embodiment and Flow,' he explained that 'we have to understand that cognition is not cold calculation. It's always got an affective, motivational, emotional aspect. All cognition is about caring or not caring about taking a risk which has affective consequences.'

His sixth E, *exaptation*, is the one that may be most relevant to how we can bring back useful information from psychedelic experiences. Exaptation is a term from evolutionary biology, describing the process whereby features of an organism acquire functions they weren't originally adapted for. One good example is the tongue, which allows us to manipulate food but was exapted to allow us to speak, as well. A bird already has feathers to fly, so nature helped it use those same feathers for showing off to potential mates. Vervaeke argues that when we're talking about concepts, 'You're basically taking the same machinery you use for moving around physical space and you're exapting it.'[34]

A practice like Tai Chi, which can improve our physical balance, can often give us a more balanced view of people or help us flow between different ideas more fluently. Practices like mindfulness that allow us to decenter and accept our experience can help us listen to others without taking what they're saying personally, and pay closer attention to what they're *actually* saying.

Because psychedelics act as such powerful connecting agents, it may be that not only can they deepen our individual sense of *embodiment* and *embeddedness* in reality, but in turn, they can help us find new ways to *enact* changes and *extend* our cognition through one another and our technology. To do this collectively, we need to have a shared reality and shared stories, a challenge I'll explore in the next chapter.

When we manage to work together well, human beings have a tremendous ability to exapt information from one domain and apply it to another. This is why the way we move through the psychedelic experience can be applied to how we move through the times we're living in. Cognitive science suggests this is not only possible, but a core aspect of how we know the world and solve problems.

As well as taking what we're learning from psychedelic science, philosophy, and indigenous cosmologies, and applying them to practical problems, it may be that we can also use them to enter into a wider distributed cognition with other intelligences. This means taking seriously the possibility that the entities that appear in psychedelic experiences are in some way real, and that the experiences we have with them may contain wisdom. I was about to encounter this firsthand.

The Teaching Presence

Lying on the chair, waiting to be dosed, a familiar feeling spread through my chest. Itchy anticipation mingled with a strange sense of peace. Peace, because there was nothing left to do. I was as ready as I'd ever be, and all that was left to do was wait. I focused on my breath. I observed the sensations in my body. What was I noticing? Anticipation in my chest. Anxiety in my belly. My legs felt more solid, so I focused on them more to ground myself. I remembered the small wooden owl I'd brought with me as a symbol of guidance and guardianship; it now sat on the altar opposite the chair. Knowing it was there helped more than I thought it would.

'Minute five. Intensity rating.'

'One,' I replied.

A cold sensation filled my left arm. A huge rush of energy – and my body lifted, burned, scrambled. Geometric patterns exploded from nowhere – a chaos of color, dimensions of meaning – and I couldn't hold on to anything. Entities crowded my periphery. They were brightly clad in neon reds and blues, sneering with the wolf eyes of a carnival trickster. They kept their distance as I searched for an anchor, something coherent to interact with. My heart was slamming in my chest.

A vast cavern appeared in the center of my vision. A geometric orb of breathtaking complexity revealed itself and spun on its axis. I felt a strong presence, what I would come to call the Teaching Presence of the DMT. It asked me what I was looking for, and my mind scrambled to find an answer. 'Truth,' I answered. It challenged me, not with anger, but it was certainly not taking any bullshit. The carnival entities squeezed past me in hidden dimensions beyond my vision. I dug deep to find a better answer. 'Novelty,' I said.

The scene shifted instantly. The orb vanished. I started to shake intensely. During this time, Lisa was noting down in detail every movement. Later, she would ask me about the shaking and I would have trouble answering exactly what the shaking meant. Most of it was happening in my jaw, where I hold a lot of tension. My teeth chattered wildly, but I didn't have a clear sense of how it related to what I was experiencing. The Teaching Presence told me I didn't need to see aliens. I needed to feel my emotions.

I felt a surge of anguish. I was shown the many ways I suppress what I'm feeling day to day; how I numb out, avoid, run away; the fear I have of truly experiencing the breadth of my own sensitivity. This was a familiar lesson, one I'd had before in other experiences. I felt almost weary of having to go back to it again. But this weariness was short-lived, because the DMT had other plans. This was a sideshow, a preamble for what lay ahead.

The DMT revealed the most profound unresolved baggage I was carrying. The Teaching Presence made clear that it was the most significant open wound in my life – what I've been avoiding without knowing I was avoiding it. It was affecting every aspect of my existence: my trust in other people, the intimacy in my relationships, my self-esteem. I was about to have my world ripped open.

What happened next was one of the most significant and transformative aspects of the psychedelic experience. Maria Papaspyrou put it best when we spoke: Psychedelics go into the depths of our psyche to reveal what is hidden. And this inner process of change was not what I had been expecting.[35]

Many months after the trial, I asked Lisa what had most surprised her about it, and she shared that the number of therapeutic or deeply personal experiences participants reported had surprised her. I felt similarly, especially because the whole lore around DMT is one of metaphysical exploration. This idea is captured best by Terence McKenna's vision of what it means to be a psychedelic explorer:

> We each must become like fishermen, and go out on to the dark ocean of mind, and let our nets down into that sea. And what you're after is not some behemoth that will tear through your nets, follow them and drag you in your little boat, you know, into the abyss, nor are what we're looking for a bunch of sardines that can slip through your net and disappear...

> What we are looking for are middle-size ideas, that are not so small that they are trivial, and not so large that they're incomprehensible. Middle-size ideas we can wrestle into our boat and take back to the folks on shore, and have fish dinner. And every one of us when we go into the psychedelic state, this is what we should be looking for. It's not for your elucidation, it's not part of your self-directed psychotherapy. You are an explorer, and you represent our species, and the greatest good you can do is to bring back a new idea, because our world is in danger by the absence of good ideas. Our world is in crisis because of the absence of consciousness.[36]

I have always been inspired by the idea of bringing back useful ideas from a psychedelic experience. However, what I experienced during my encounter with the Teaching Presence would lead me to question and reject McKenna's split between the personal and the metaphysical. I now believe they are intricately woven together. When Erik Davis and I spoke, he told me that he believes McKenna's greatest blind spot was the importance of inner work - and despite his genius, he had a penchant for spiritual bypassing, the process of using spiritual ideas to avoid our own emotions and our pain. It is a form of avoidance, a key psychological defense we all have, but it can be particularly damaging with spiritual practices because they can convince us that we don't really need to go through the pain of being human and confront the messiness and complexity of our lives.

I was about to get a full dose of this understanding, because the Teaching Presence of the DMT was leading me not out into hyperspace, or to entities who could give me brilliant insights, but right into the depths of my own life. It revealed to me a situation related to my mother that has been a significant barrier to our relationship since I was a teenager - a situation that has created a rift in our family and in my relationship with her.

A wave of grief and shock punched me in the gut. I didn't think about this matter much in my life. I didn't want to. I talked to my mother, but we never talked about *this*. Our silence around it had become an unspoken agreement. Letters had been written; conversations had been attempted. But I could feel in that moment, with the Teaching Presence, how much I loved her and how much I missed being closer to her.

The Teaching Presence became firm and clear. It told me that by avoiding my pain around the situation, I had cut myself off. I was in an emotional limbo. If I wanted novelty, it told me, this unresolved baggage was what was keeping me from it. I started to weep as I saw how profoundly the situation had shaped me;

how difficult I found it to trust people; how alone I sometimes felt, even when surrounded by those who love me; how utterly helpless and hopeless I felt in relation to this.

As I wept, the Teaching Presence told me that grief was healthy, but what I really had to do to move forward was to decide who I was going to be in relation to my mother and the situation – not fixing it or trying to change it, but coming into 'right relationship' with her.

As the dose wore off, I was left with a sense of grim acceptance and a familiar feeling of having seen something that, while painful, was necessary. I had absolutely no idea what to do. I was left with complexity, and a clear knowledge that only I could decide who I wanted to be in relation to what I'd experienced, and to the situation itself. But despite the swirling intensity of emotion, I felt lighter and clearer; more alive and connected to the world. And just as the dose was about to wear off completely, the Teaching Presence showed me something else: a dim scene of multiple portals that seemed to lead to other layers of this reality. It told me that DMT acts as a nexus to multiple dimensions. We can travel anywhere we want, it said, but we have to take our emotional baggage with us.

Closing Comments

What are we to make of experiences like these? Not just my own personal experiences, but thousands like it, which we've seen in clinical trials. Beyond the research being done, millions of encounters like the one I recounted are being had among small groups of friends and religious rituals and ceremonies around the world.

What they do, and why they are so profoundly useful for giving us a new perspective on collective problems, is that they take us out of our existing frames of reference. They may serve to acquaint us with a completely different reality or, as in my case, with what we aren't acknowledging within ourselves.

As I argued earlier in the chapter, being able to step outside of our existing frame of reference can be profoundly transformative. But to do that, we need to understand what our existing frame actually is. And while it's unique for each of us, if you live on planet Earth right now, you share aspects of your view of reality with many others. We are all changed and molded by our environment, and right now, our collective environment is in trouble.

Our vast, interconnected global civilization is running out of road. We are facing so many complex problems at once that we have no idea how to solve them. And yet, just as I was the only person who could solve my own personal issue revealed in the dosing, collectively speaking, we are the only ones who can solve the issues that affect all of us: the environmental crisis, the mental-health crisis, rising wealth inequality, racial inequality, increasing political polarization – and the list goes on, and on, and on.

In the next chapter, we are going to look at these problems more deeply and see how they create a particular frame on reality – and how what we're learning from psychedelics could help us make sense of what's going on in completely new ways.

Chapter Two

The Big Crisis

There are stories hidden deep in your bones. Stories that were lived before you were born by the thousands of women and men who laughed, danced, fought, and died so that we could be here now. At countless points during the long history of humanity, our ancestors were faced with a crisis. A moment of decision.

Perhaps they stood on a mountain and looked out at a valley no one had ever crossed. A valley that had to be traversed because staying in place was no longer an option. And as the first of them made that first step, she might have been aware that she and her tribe ran the risk of not surviving the journey. Leaving what we know brings us into a new world pregnant with potential and wild with danger. It's easy to lose our bearings – to forget not just where we're going, but who we are.

In the deepest parts of ourselves, we know this journey. It is the journey into the liminal, the space between worlds. We experience it when we quit our jobs and start a new career; when we enter a new relationship; when a pandemic freezes the world and we don't know what's coming next. It is the journey long associated with altered states and psychedelic experiences – a process of traversing the underworld. This is the collective journey we are on right now as a species. We are walking out of a past we know can no longer sustain us and into a future that has yet to be born.

And while it can be disorienting and scary, it's also a time full of possibility. In ancient Greece, there were two words for time: *chronos* and *kairos*. Chronos, from which the word 'chronological' is derived, relates to linear time. Kairos is different: a kind of immediate, shifting moment in which a new opening appears that can be grasped, as long as you're paying attention.

We are living in kairos. A culture that defines itself on striving its way to the future has been forced into the valley of the unknown by the recognition that the way we're living cannot go on. This chapter is about making sense of the unique situation we find ourselves in as a species right now. It's about how psychedelic experiences and other wisdom practices can help us traverse the uncertainty and chaos around us, by opening us up to new ways of seeing and being. It's about how a series of crises – in economics, culture, media, myth-making, and spirituality – are converging all at once, and why we need new frames on all of them.

The Crises of Our Times

Our frames on reality are formed in large part by the story of the world that our culture tells us. Every culture has stories it lives by. For the last few hundred years, the dominant story in much of the world has been that of material progress.

We live in an age of certainties falling apart, and we have no idea which path to take. The kind of progress that has led developed nations to enjoy a higher standard of living than our ancestors is also destroying the planet. The story of many secular countries is that of continuous consumption and growth, what I call 'consumer culture.'

And while that story tells us what to do, it doesn't tell us why. What, precisely, are we trying to progress toward? An unending age of science and reason? The colonization of Mars? The ensuring of social justice and a fair economic system? Eternal life in a corporate metaverse? There are as many answers as there are people, and if there is a myth for the times, it is the Biblical story of the Tower of Babel – of not being able to understand one another or come together toward a shared purpose, even when faced with extinction.

As my friend Ivo Mensch has put it, 'We're collectively living a life that no longer exists.'[1] The road we're walking leads to environmental collapse, increasing rates of political polarization, and a breakdown of trust in almost every area of society. This fragmentation runs within us and between us.

Between us, because we are increasingly spending time on social networks that allow us to live in our own realities, with our social feeds giving us information we already agree with, or information that will enrage us and keep us glued to the platform. As a result, it's increasingly hard to cohere around shared stories and understandings.

But perhaps this doesn't matter. Perhaps, as Steven Pinker has suggested, we're actually living in the safest and most exciting time in human history.[2] Maybe we'll be able to innovate our way out of the ecological crisis and drug our way out of the mental-health crisis. Our technological advances and our existing economic systems have lifted millions out of poverty. We enjoy a standard of living our ancestors couldn't have dreamed of, enabled by awe-inspiring technology and a more compassionate, emotionally aware culture than we had a hundred years ago.

But the situation is complex. Because what this perspective misses is that, while we've enjoyed material progress, it has come at the cost of meaning. John Vervaeke has suggested that we're living through 'a meaning crisis.'[3] While its roots are complex and go deep into our history, the result is that many people feel increasingly detached from the world, and from a deeper sense of purpose. Science has told a story that the world is a dead machine, and human consciousness a random fluke in an uncaring universe.

The global decline of religions means that in many cultures, this secular narrative is dominant. Nietzsche talked about the 'death of God' and warned that once you remove the deeper architecture of spiritual belief from society, you leave a vacuum that will inevitably be filled by things you didn't expect. With no convincing religious frameworks to fill the gap of purpose and give us a deeper sense of coherence, late-stage capitalist cultures lack, at a foundational level, a connection to something beyond themselves; this generates what Vervaeke has called 'cultural narcissism.'

As a result, our politics and ideologies have filled the void of religion, giving many a sense of purpose and meaning. However, this has also brought chaos. Research shows that voter turnout is declining globally, particularly among the young.[4] Increasingly, old political divisions of left and right don't capture people's concerns, and the political landscape in the West has splintered into what Peter Limberg and Conor Barnes have termed 'memetic tribes.'[5] These are groups that share a common set of values and fears – for example, around intersectionality, or cryptocurrency, or conspiracy theory – but don't necessarily identify with a political party. As they battle for narrative supremacy, they're creating a 'culture war' that is manipulated by governments and corporations for political gain and deepens our fragmentation and alienation from one another.

Name a school of thought that ends in 'ism' and you'll find a memetic tribe with its own unique creed. Capitalism. Marxism. Post-colonialism. Libertarianism. Neoliberalism. Political creed increasingly stands in place of religious creed, but neither brings us toward a genuine connection to something that takes us outside ourselves and allows us to make new frames.

COVID-19 accelerated these trends. Stuck indoors and trapped online, many of us got lost in rabbit holes of magical thinking and conspiracy theory. More than ever, social networks incentivized political tribalism and outrage as we argued over vaccines. The disinformation, propaganda, and narcissism that pervade the online world went into hyperdrive. Desperate for reliable information and understandingly distrustful of the traditional media, millions turned to alternative media sources to make sense of what was going on.

But nobody regulates the alternatives, which face their own perverse incentive structures. As a result, knowing which sources to trust is becoming harder and harder. Nuanced, compassionate conversations about important social topics like race or gender are harder still. Facts, whether about vaccines or politics, were swept into a tornado of mythic dreaming and semi-religious fantasies spread across countless ideological tribes fighting for narrative control.

This is the culture into which psychedelic drugs are going mainstream. A culture of desperation and chaos, swinging wildly between narcissism and nihilism, and running out of time.

Welcome to the Big Crisis.

An Initiation Process

The word 'crisis' has its roots in the Greek *krinein*, 'to decide.' It's an inflection point at which we must decide whether to evolve or die. It's a moment when everything seems hopelessly fragmented, and we must somehow find a way to make the broken whole again.

The time we're living through is similar to how some shamans describe the process of initiation. Tungusic shamans in Siberia have reported being dismembered by spirits who counted their bones to determine if they had what was required.[6] On the other side of the world in the Amazon, shaman Davi Kopenawa, in his book *The Falling Sky: Words of a Yanomami Shaman*, reports being tested by the *xapiri*, spirits with which the Yanomami commune through a mixture of DMT snuff and tobacco.

Kopenawa writes:

> *When they arrive they also hurt you and cut up your body. They divide your torso, your lower body, and your head, they*

sever your tongue and throw it far away for it only speaks ghost talk. They pull out your teeth, considering them dirty and full of cavities. They get rid of your guts, full of residues of game, which disgusts them. Then they replace all that with the images of their own tongues, teeth and entrails. This is how they put us to the test. This is what happened to me and I was truly scared.[7]

Being truly scared was a regular experience on the DMT trial for many of us. Speaking to other participants once the trial had ended, I noticed a common theme that one of the most intense moments was waiting for the DMT to be injected. Usually, psychedelic experiences come on gradually, peak, and then decline. In our case, reality went from zero to a thousand miles an hour within a few seconds.

As Lisa Luan told me while she was writing up the results, all of our heart rates shot up for those first few minutes, then leveled out. The first dosing I reported in the previous chapter was the lowest dose I would receive on the trial, and as I prepared for the next dose, I was keenly aware that I was on a process that would be as much about resilience as it would be about exploring the unknown.

In shamanic cultures that don't use psychedelics, practices that involve extreme duress on the body, like trance dancing or hanging from hooks, are often utilized to achieve altered states. In many of these practices, there is a common theme: To access powerful new ways of seeing the world, first we have to be pressured and tested. In some way, we must willingly lean into the chaos of having our world flipped upside down, and stay resilient in the face of it.

That is one way to understanding what's happening to us collectively right now. We are living in what Ken Wilber has called 'a post-truth world.'[8] It is one in which we are challenging all fixed notions of who we are and what truth is. Leading theories in the humanities argue that the world is constructed by relationships of language and power, and that concepts like gender and race are socially constructed. The idea that there is an underlying, essential truth to reality is rejected by these theories. Similarly, scientific materialism, which argues that the only thing that's real is matter and that consciousness is simply a byproduct, leaves us with a cultural story with very flimsy foundations. We can go through a process of fragmentation if there's a larger reality to fragment into that can provide an underlying sense of coherence to our lives. Otherwise, we are left with a perpetual deconstruction of anything certain that is psychologically agonizing for many people. Sociologist Aaron Antonovsky has proposed that a 'sense of coherence' is crucial for our well-being and a sense of meaning in our lives, as it helps us contextualize our existence within a wider scheme of meaning.[9]

A life without coherence can feel hopeless and terrifying. But being torn to pieces can be a profoundly initiatory experience. Initiation is common in myths and stories around the world, and we can reframe this collective process as an initiation we're all going through that psychedelic experiences can help us make sense of.

Initiation is a common feature of the psychedelic experience; it's the process whereby we go through experiences that challenge our previous conception of who we were and force us to become someone more than we were before. It is an experience of breaking our old frame on reality, and building a new one that can hold more of the world around us.

A significant number of people report that the psychedelic experience is a 'rebirth.' But in order for our perception to be reborn, we have to let our old views die. Letting our cultural assumptions die can be a revolutionary act. It is an idea I first heard from Stephen Jenkinson, a writer and storyteller known for his bestseller *Die Wise: A Manifesto for Sanity and Soul*, which chronicled his time as a death and dying counselor. When we spoke, he pointed out that a culture founded on the story of infinite growth and progress has no place for endings and doesn't know how to handle them.

'Imagine you really love somebody who's dying. What is the etiquette of your relationship with them? How do you conduct yourself as if what's happening is happening? That becomes the question for a dying person in your household and it becomes the same question for a time like now, where we get to glimpse how utterly exhausted our acquisitive way of doing things has become. So, what do you do? The answer to my mind is you bear faithful witness to the realities of the day. You orchestrate yourself in the presence of the dying, not despite it, not to overcome it, not to pretend it's not there for the sake of mutual comfort, but to risk it all, to risk the relationship, the friendship, whatever it is. And fail to conspire with the idea that there's more time. There is no more time.'[10]

We are running out of time, and we can't acquire any more. The world is heating up. Our geopolitical situation is intensifying, and the global order is fraying. And our normal response in this situation is to strive – to impose our will on it to force change.

We see this response in our alarmism, in our penchant to wage war on anything we view as a threat: the war on drugs, the war on climate, the war on cancer. But death and transformation can't be reasoned with, or attacked away. Instead, they need to be placed within a bigger picture.

Myth and Reality

Many indigenous and non-Western cultures that use psychedelics do so within a coherent cosmology and mythology, which is often very explicitly connected to the kind of humility that is required as we consider our place within the bigger picture. It allows for a reciprocal and respectful relationship to place, to the land itself. As psychedelic anthropologist Gabriel Amezcua points out, it is vitally important that we don't fall into a colonial mentality by assuming that indigenous people represent a homogeneous group or have a mystical know-how that Westerners can acquire to fix our own issues.[11] Psychedelic-using indigenous groups are as complex and diverse as humans anywhere, and the unique solutions they've found to address their problems are context-specific and may not necessarily work for other cultures.

However, this particular quality of having coherent myths to live by is something we would do well to pay attention to and learn from in the West. Myths remind us of what it means to be human, where we're going, and why. As late-stage capitalist culture fragments, and we look for our own roots, do we find they go deep enough to give us that same coherence?

I don't believe they do. However, deeply encoded in our psyches, we may have the instructions we need for how to be human in the in-between time we're living in. To find them, we have to know how to reach deeply into our own roots and wrap our fingers around the stories in our souls. We have to learn to speak the language of the underworld: a grammar of symbols and archetypes, intuition and instant knowing, dreams and secrets, and great dramas played out over eons beneath the surface of things.

Psychedelics can teach us this language. But it's a dangerous language. It's the language of transformation, of changing who we are and how we conceive of ourselves. It is a language that burns you, that shreds away who you thought you were and presents you with something much vaster and more beautiful – and not at all what you expected.

Complexity and the Meta-Crisis

Aldous Huxley described the psychedelic experience as opening up the 'reducing valve' that limits our perception. In doing so, we also 'open the doors of perception' and see much more of the world than we did before.[12] One way to describe this is that psychedelics open us up to complexity. This may be one of the most significant ways they can help us make sense of the times. To see why, we need to take a deeper look at what *kinds* of problems we're facing.

The Big Crisis is often called the 'meta-crisis.' Both are terms for something so all-encompassing it can be hard to really make sense of it. Jonathan Rowson, a former chess grandmaster and philosopher, wrote a seminal essay on the meta-crisis called 'Tasting the Pickle: Ten flavours of meta-crisis and the appetite for a new civilisation.' In it, he argues that understanding it isn't just conceptual, but also aesthetic. We have to *feel* it. He also argues that even defining it as a singular crisis obscures it. Instead, it may be more accurate to conceive of it as multiple overlapping crises. He writes:

> Once you take the idea of meta-crises seriously and start looking at them closely, it seems we are caught up in something oceanic in its depth and range, and plural. The idea of trying to define the meta-crisis as if it could be encapsulated as a single notion and conceptually conquered is a kind of trap. I have come to think it helps to distinguish between different features of an experience that ultimately amount to the same underlying process. In fact, that's how I see the meta-crises writ large: they are the underlying processes causing us to gradually lose our bearings in the world.[13]

Author and philosopher Tomas Björkman sees the meta-crisis as a developmental challenge in many ways. It's a moment in history when we're being asked to collectively develop and evolve not just where we're going, but *who we are*. Björkman has written a lot about the concept of *bildung*, the idea that we are continuously learning and developing our intellectual, spiritual, and moral capacities as adults. For Björkman, this bifurcation point doesn't run through society only, but through each of us. We're being asked to develop – to change not just our behaviors, but the values driving them.[14]

As Rowson and Björkman both point out, the Big Crisis consists of many overlapping crises: the mental-health crisis, the environmental crisis, the crisis of trust in our institutions, and many more. And this makes it not just complicated, but complex.

Maps of Complexity

To see why that's important, we can look at what the word 'complex' actually means. It can help to compare it with the word it most often gets confused with: 'complicated.' If something is complicated, it has many different parts that interact with one another in specific and limited ways. A car engine is a good example. If, like me, you need a YouTube tutorial to change a car battery, then you don't really know how a car engine works, and how the different parts connect. Figuring that out may feel daunting, but given enough time and motivation, you could. You just need some schematics, a manual, or a good teacher.

Similarly, a grandfather clock is extraordinarily complicated, with hundreds of tiny parts moving in just the right way, with just the right friction to keep the clock going. If one part breaks, the system breaks down. To fix it, you just open the clock and identify which part or parts aren't working, and replace them.

The car engine and the grandfather clock are intricate machines, but they follow linear rules: Change something in one place, and you can predict what will happen in another. They are both examples of a closed system. It's all wrapped up in your bonnet or in the clock casing, and their parts only interact with one another in limited ways. When something goes wrong in a complicated system, like a broken clock, it can be very hard to fix. It requires a lot of thought, clever processes, and plenty of trial and error. However, once you figure out the problem, you can apply some kind of rule and fix it. That might entail a replacement part in a clock, an algorithm in a piece of software, or a new workflow in your production line.

Complexity is something very different. In his book *Hidden Order: How Adaptation Builds Complexity*, mathematician and psychologist John H. Holland describes how, whether we're talking about an ant colony or a city's economy, we see the same patterns and behaviors arising from what he and others have called 'complex adaptive systems.'[15] Holland is famous for his theories on how genetics adapt and change, and for a time, he worked with the Santa Fe Institute, an organization set up to study complexity.

Complex systems are open, meaning they respond to other systems and inputs. They are made up of many different parts that all interact with one another in a way that leads to non-linear outcomes; that is, change something in one place and you can't predict what's going to happen. An easy way to think of a complex system is that it's alive, even if it isn't in fact a living organism. Your body is a complex system, with millions of cells each individually doing their thing at different levels of coordination. Some of them form a liver. Some of them form your spinal cord. All of those billions of cells interacting together, and responding to their environment, create an organism called you. But you are much more than the sum of your parts.

A good way to think of the difference between the two types of systems is by imagining a grandfather clock sitting on top of an anthill. The clock is complicated; the ant colony is complex. The clock is a closed and linear system: The way it behaves is the result of all its individual components and nothing else. The ant colony is open and non-linear: It isn't closed off; rather, it's in a relationship with its environment and is constantly adapting and changing. The colony itself has a higher intelligence than any individual ant; it's the millions of interactions *among* the ants that create the higher adaptive intelligence.

Another quality of complex adaptive systems is diversity; the rainforest is teeming with plants and animals in the same way New York City is teeming with different kinds of people, stores, and amusements. That diversity creates a kind of resilience; if you remove a single ant in the colony, the ant will soon die, but the colony will adapt. If you remove a piece of the clock, the whole device might stop working immediately. Complicated systems may be impressive, but complex systems are sublime and beautiful.

Perhaps the most significant and mysterious quality of complex systems is that they generate something known as *emergence*. Emergence is the process whereby new phenomena grow out of the interactions of complex systems. We could see emergence as a kind of creative novelty, in which new things come about not because they are designed, but because they are birthed by the process of reality itself.

Nature didn't sit down and design anteaters over a cup of coffee. Anteaters emerged through the complex interactions of evolution. Likewise, while social ideologies and artistic movements may sometimes be attributed to one person, these also emerge from the complexities of human interaction and the exchange of cultural ideas. As Harvard evolutionary psychologist Joseph Henrich has pointed out in *The Secret of our Success: How Culture Is Driving Human Evolution, Domesticating Our Species, and Making Us Smarter*,[16] it's humans' ability to constantly adapt and combine cultural ideas that makes us so unique among other animals.

New ideas – a novel type of basket, economic system, religion, etc. – are tested out, blended together, and prodded until the most adaptive survive. Then, the process repeats. However, there is no single culture or person designing *who we are* in the way a potter makes a pot; it's an aspect of evolution and emergence, and it's happening around you and through you right now.

You are participating in culture, and culture is participating through you. In an ant colony, new, unexpected behaviors emerge from the astonishingly beautiful and unpredictable interactions of all the individual ants; the intelligence and responsiveness of the whole colony is an emergent property of the system, but no individual ant is intelligent. Individual humans are intelligent, but we still create emergent phenomena, from clothing trends to new systems of thought, through our interactions with one another. With the advent of the Internet, this process is now happening so fast that many of us feel stunned by it.

Your brain is a complex adaptive system, and what we're learning from psychedelic science can point us toward a new way of approaching complexity. One of the big surprises in the early brain-imaging studies conducted by Imperial College was that, in some ways, psilocybin leads to *less* activity in the brain, not more. Specifically, psilocybin reduces activity in an area of the brain called the default mode network (DMN).

Carhart-Harris and others have theorized that this is the area linked to the narrating 'I' in your head. In depressed patients, the DMN is overactive, which may partially account for ruminating or hyper-critical thoughts. As Carhart-Harris argues, the default mode network 'serves as a central orchestrator or conductor of global brain function... engaged during higher-level, metacognitive operations such as self-reflection (Qin and Northoff, 2011), theory-of-mind (Spreng and Grady, 2010) and mental time-travel (Buckner and Carroll, 2007) – functions which may be exclusive to humans.'[17]

When we reduce the input from this narrator, through meditation or psychedelic experiences, different parts of the brain can speak to one another. The whole complex system starts to interact in new ways. We can make new neuronal connections, come up with new ideas, and with the right spiritual or therapeutic guidance, heal our trauma. What is crucial here is that this is *not done through an act of will*. A change in frame comes about not just through imagining a new one, but letting go of your old one.

In the same way, we can't change the dynamics underlying the Big Crisis if we use our old cultural frames, which tell us that changing things is a matter of controlling variables. We can't just add in new laws or cultural norms, and expect that's going to change everything. And yet, many of our policies and even our education system are based on trying to solve complex problems with complicated solutions.

Psychedelics can give us a lived experience of navigating complexity. So much healing in psychedelic therapy and ceremonies happens when we accept the complexity of how we feel. When we go through the constantly changing, interwoven layers of our experience to find the truth. When we accept who we are, how we feel, and what happened to us.

This process is echoed in many spiritual traditions, which hold that one of the purest expressions of our divine nature is allowing ourselves to flow from a deeper essence within ourselves – not second-guessing, not trying to control who we are, but simply letting go into the truth of what is. It is core to the Taoist concept of *wu wei*: a flowing, effortless action that is in tune with reality. And it is an attitude thrown into sharp relief with many policy decisions and cultural conversations that are attempting to respond to our complex problems.

Too often, we are stuck, pushing *against* complexity, pushing against ourselves. Learning how to flow with *what is* on a cultural and individual level is absolutely essential for navigating the Big Crisis and a psychedelic experience. To be able to change our ruinous trajectory as a species, we have to learn how to first adapt to the increased complexity brought about by our technology, and then consciously change our values, systems, and politics to be in alignment with what we're experiencing. It is strikingly similar to the way in which, to attain insights in a psychedelic experience, we must flow with the complexity of what's happening and adapt our perspective to the new truths we're encountering.

The first step in that is realizing how useless it is to try and explain everything through an 'ism' that tries to squeeze a complex reality into a small conceptual frame. Whatever ism you find, it will contain this error, suggesting that 'if only' we changed one aspect (like taking a piece out of a grandfather clock), everything would work better. Whether that's a neat political ideology, conspiracy theory, or sociological theory – unless it brings us into a lived experience of contradiction, complexity, and emergence, it's a comforting fantasy and will only ever be partially true in very limited situations.

Isms will get us absolutely nowhere. We need new ways of seeing and being that honor the complex, changing, dynamic reality we're part of. That is the first step in taking action to change its course. Many people call this 'systems change,' but Nora Bateson, a renowned systems theorist and philosopher, often encourages people to move away from the idea of systems change and toward *systems learning*. Complex systems, as Holland and others have pointed out, don't just change; they learn. They evolve based on past patterns, with a kind of intelligence and adaptability. And so, when we're trying to make sense of what's going on, we have to first recognize that what we're trying to make sense of is *changing as we're trying to make sense of it.*

Zen masters used to instruct samurai to be 'like a barrel of water.' When you pierce a barrel, the water flows out without hesitation. It doesn't stop to think about what model to use, or try to analyze the situation. A moment of hesitation could mean death for a samurai, so they cultivated an attitude of 'no-mind': remaining awake and present to the here and now, not lost in a story of the past or a model of the future. I have found this attitude to be essential for navigating the rush of emotional and sensory complexity that accompanies a psychedelic experience.

At the same time, I've found psychological theories, spiritual traditions, and cognitive science hugely useful for integrating psychedelic experiences in the long term. In a similar way, when we're making sense of the world, we need some kind of map: a model that we can refer back to that captures at least some of what we're trying to look at. Polish American philosopher Alfred Korzybski famously popularized the idea that 'the map is not the territory.'[18] A map is an attempt to create a model of reality that's *good enough* to help us make sense and take meaningful action. If we're trying to map the Big Crisis, it means drawing a map on sand - a fleeting snapshot of a piece of the dynamic, mind-bogglingly complex social reality we're all living in. This is not a map to hold just in our heads, but in our hearts.

Hyperobjects

The first stage of building this kind of map is returning to a sense of openness and humility, recognizing that many of the problems we're facing are just too big to grasp, and being OK with not knowing... sitting with the uncertainty and the pain this sometimes entails.

One way to describe this is through the concept of 'hyperobjects,' a term coined by Timothy Morton in his book *Hyperobjects: Philosophy and Ecology After the End of the World*.[19] A hyperobject is something that's all around you, but that you can't touch or grasp in its entirety. Global warming is a hyperobject. Evolution is a hyperobject. The sum total of all the plastic in the world is a hyperobject. Morten argues that if you come across a piece of plastic litter, you're seeing a 'local manifestation' of all the plastic in the world. That single piece is part of the whole, but in itself isn't 'all the plastic.' Similarly, you can point to a platypus and say, 'Look, this is evolution!' And yes, the platypus may be an example of evolution, but evolution itself is so vast and interweaving that it will always escape our grasp. Evolution is moving through and around you at all times. Climate change is something you are embedded in and affected by that you can never touch. So, where is it? Everywhere, and nowhere.

The Big Crisis is a hyperobject made of other hyperobjects. One of the things that psychedelics can help us do is *feel* hyperobjects; to be confronted with a profound level of complexity and intricacy without being able to process it fully, but nevertheless achieving a better sense of wholeness, either in ourselves or around us. However, what can seem so clear from a zoomed-out perspective in an altered state becomes harder to integrate into our day-to-day lives. And so, not only do we need to hold any conceptual map of the Big Crisis very lightly, recognizing that our view of the world isn't the same as the world itself, but we also need to stay aware of *the way we're reading it.*

A Bicameral Problem

If it seems difficult to look around and see the complexity, rather than lots of individual parts, you aren't alone. It may be due to the unique two-hemisphere structure of our brains, which actually changes how we see reality. The idea that the left and right hemispheres of our brain perceive the world differently was popular in the 1980s and resulted in an oversimplified notion that the left hemisphere is logical and the right is creative. This view was far too simplistic and led to the idea of the 'bicameral brain' being largely discounted by neuroscientists. However, as psychiatrist Iain McGilchrist points out in his bestselling books *The Master and His Emissary: The Divided Brain and the Making of the Western World*[20] and *The Matter with Things: Our Brains, Our Delusions and the Unmaking of the World*,[21] while it's clear both hemispheres are involved in many aspects of our perception and a more complex view of them is required, we have thrown the baby out with the bathwater by ignoring reams of evidence suggesting that our two hemispheres do in fact perceive the world in very different ways.

McGilchrist argues that is fundamentally important to how we understand not just ourselves, but our whole civilization. We will see echoes of the model McGilchrist presents throughout this book in many different forms. In short, he draws on a huge body of research to argue that we are living in a culture defined largely by the left hemisphere's view of the world. In broad terms, the left hemisphere is focused on what we already know. Its attention to detail is narrow, and it perceives bodies as an assemblage of individual parts. It is a master of abstraction, constantly breaking things (and people) into their component parts. It is the perspective of the predator: hyper-focused and haughtily certain about the truth of its own perception. It has clarity and the power to manipulate things that are already known, explicit, and general. But it also behaves like a closed system and finds it hard to conceive of things it doesn't know. Despite this, it is certain that it definitely *does* know. It is important to point out that both hemispheres are involved in reason and emotion, so the old model of the left hemisphere as the coldly rational side is inaccurate. It does, however, have a kind of coldness to its perception. McGilchrist describes the left hemisphere's gestalt as 'perfection bought at the price of emptiness.'

The right hemisphere, McGilchrist points out, is more responsible for interpreting what we *don't know*. Its focus is sustained, broad, open, vigilant, and alert. It sees things in context and understands implicit meaning. It thrives on metaphor and understands body language. In a particularly cogent example, McGilchrist draws on case studies of artists who have had a stroke in the left hemisphere (in which case the right hemisphere becomes more dominant). Very often, their art becomes more creative, more unusual, bolder. The right hemisphere lives in an embodied world – a world full of individuals, not just categories. Its universe is never fully graspable, never perfectly known. It has a broad, contextual understanding and sees people as people rather than assembled parts.

It's striking how different reality seems when the left hemisphere dominates. McGilchrist presents a case study of a man who had a stroke in his right hemisphere (resulting in the left hemisphere picking up the cognitive slack). When the doctor came in and asked, 'How are you feeling?' the man responded without any irony, 'With my hands.' The left hemisphere doesn't get metaphor.

We've been drifting further to the left hemisphere's point of view: hyper-rational, reductionist, and certain of our own ideologies. The answer is not to swing the pendulum the other way and call for a right hemisphere-dominated society. Instead, it is to find balance and harmony between the two perspectives.

When we spoke, I asked McGilchrist what he makes of the fact that so many people report a more holistic, interconnected perspective after taking psychedelics. It seems reasonable to hypothesize that psychedelics can bring us into a more right-hemisphere perspective. However, he was quick to point out that we have no research to suggest this, and in fact, some LSD research from the 1960s which would seem to negate the idea that psychedelics interact in a meaningful way with the right hemisphere within this context.[22,23]

There is one potential link between psychedelics and bicamerality, which comes from the research of American psychologist Julian Jaynes. In his book *The Origin of Consciousness in the Breakdown of the Bicameral Mind*,[24] he noted (according to Rick Strassman's summary) that 'For millennia, those existing in small tribal groups perceived an "outside" voice advising, admonishing, and otherwise guiding individuals who faced various decision points. The spoken voice served as the prototype of self-consciousness and introspection. Jaynes theorizes that this voice was the subjective experience of communication between the brain's right and left cerebral hemispheres.'[25]

For Jaynes, this interaction between the hemispheres was eventually lost as civilization grew, and prophets were the last examples of people who followed that voice. In *DMT and the Soul of Prophecy*, Strassman inquires as to whether ancient prophecy might have been mediated by endogenous DMT if it is indeed a result of the bicameral mind. Whether or not DMT or other psychedelics play a role in helping our hemispheres talk to one another is unknown at this stage, and it's important to point out we don't have any evidence confirming this. However, what is striking is the prevalence of a 'guiding voice' in many people's experiences – the same kind of guiding voice that Strassman points out was a hallmark of the prophetic experience for thousands of years and the DMT experience today.

A World of Information

So far, I have been building a conceptual map for use in making sense of the many overlapping crises currently going on in the world. Now, we can apply that map to looking at some of the elements of the Big Crisis that affect our daily lives most directly, and determining where psychedelics fit in.

The world has always been complex, but one of the main differences affecting our lives today is the scale and power of our technology. It is our technology, from oil and nuclear weapons to social-media algorithms and AI, that has taken our perennial problems and turned them into existential threats.

One of the leading voices in the study of how all these complex elements interact is Daniel Schmachtenberger. Schmachtenberger was a regular Rebel Wisdom guest and runs a nonprofit think tank, the Consilience Project, that tries to provide nuanced resources for helping people make sense of complexity. Along with many others, he points out that the stakes of civilizational collapse changed immediately with the invention of the atomic bomb. For the first time, we had the ability to destroy everything if we destroyed ourselves. When the Roman Empire collapsed, it was a big deal, but life went on. If our civilization goes the same way, humanity could very well end with it.

In order to act with enough wisdom to prevent this from happening, Schmachtenberger argues that we need to find new ways to coordinate at scale. However, this requires a healthy 'information ecology' in which we can be reasonably sure that the information we're sharing is true. If we can't even agree on what's going on, how can we decide which way to go?[26]

But our information ecology, which now exists mainly online, is broken. Not only are we flooded with so much information that we can't process it, and by sources we don't know if we can trust, but *the idea of truth* itself is being weaponized. It's something that individuals, governments, and corporations are twisting, molding, or creating to gain control of the narratives that lead to real-world actions.

In the summer of 2019, just months before the storming of the US Capitol on January 6, the federal Cybersecurity and Infrastructure Security Agency published a report about how foreign governments use disinformation to divide populations.[27] Ironically, the authors of the report had to choose a topic as an example that wouldn't be politically divisive within the US government, and so they chose whether or not pineapple belongs on pizza (in my view, it absolutely does). In the report, they lay out a five-stage process.

First, a foreign government will choose an issue that is already divisive, such as gender or race. Next, they create a lot of social-media accounts, with many often controlled by the same person. After this, they run a campaign to amplify and distort the conversation, playing both sides. One group gets messages saying, 'Pineapple on pizza is un-American!' while another segment might be fed with posts telling them that Republicans are trying to control their bodies by banning pineapple. In the fourth stage, foreign influencers make enough noise around an issue to get it into mainstream awareness, often through an established media organization.

Next, they bring the division from the Internet into the real world, often creating event pages and encouraging people to get out into the streets, or to a protest. As reported in the podcast 'Will Be Wild,' almost exactly this format occurred in the buildup to January 6 around the US election results. A movement that began on social media feeds then spilled into mainstream media coverage, real-world protests, and eventually, an insurrection.

It isn't just governments that use our information ecology to achieve their aims. Carl Miller is a researcher and co-founder of the Center for the Analysis of Social Media at the think tank Demos. In his 2018 book *The Death of the Gods: The New Global Power Grab*, he examines how different governments, individuals, companies, and militaries use social media and the manipulation of information to achieve their strategic goals. Miller points out that hackers now form 'a strange new kind of ruling class,' and whoever can manipulate the technology through which we make sense and connect now has the true power in society. Information itself has become the most powerful weapon of our age.[28]

This matters for how we make sense of the world, because most of us now get our news primarily through the social-media networks where this is all taking place. The networks themselves are designed not to help us find truth, but to keep us online as long as possible to serve us ads. Tristan Harris, a former Google ethicist and star of the documentary *The Social Dilemma* has revealed how social-media algorithms are specifically designed to feed people with content that sparks outrage, creating what he calls 'a race to the bottom of the brainstem' and pressing on emotional triggers. Outrage keeps us clicking longer, and it's by design that polarizing content rises to the top of our newsfeeds.[29]

So, if our social-media feeds are full of disinformation and arguments, why not just watch the TV news to make sense of what's going on? For good reason, many people no longer trust traditional media, either. David Fuller has pointed out that, as a result, we're living in an 'uncanny valley of truth seeking.'[30] On one side, we have news and academic institutions and think tanks that have formed over hundreds of years. These institutions won't often cover certain stories, and get stuck in ideological ruts. Alternative media online starts filling in the gaps that these institutions can't, or that they refuse to look at. However, alternative media outlets can quickly become captured by their own bad incentive structures and the same algorithms

Harris has pointed to. As a YouTuber, why release a nuanced, considered piece on a topic when you could get ten times the views and ad revenue by releasing content on a polarizing topic that you know your audience is going to love? As I'll demonstrate later in the book, in the online world, we can easily have our own values hijacked by those of a social-media company.

Online alternative-media sources can be excellent, but they exist in a wild west where it's very easy for people to be captured by their audience. If you get more clicks, and more money, by perpetuating outrage, the incentive structure can lead you to creating more and more content designed to inflame, enrage, and provoke. Blogger Venkatesh Rao has called this 'The Internet of Beefs'; in this structure, people with large audiences (Knights) rile up their followers (Mooks) and encourage them to promote their ideas and attack those who disagree. This creates ongoing dramas and conflicts, a total narrative warfare in which the Mooks act as pawns, driving clicks and engagement for the Knights to make the money.[31]

The result is chaos. We don't know where to go for trustworthy information, and when we try on either side of the valley, we're left wanting. The TV news says one thing. Our friends on social media say another. Articles appear on our feeds saying something completely different. We're entering a narrative war in which everyone is shouting their preferred truth, from governments and social-media influencers to click farms in Indonesia and PR companies in New York.

One example is around the lab-leak hypothesis that emerged during the COVID-19 pandemic. At the start of the pandemic, the idea that COVID leaked from a lab in Wuhan that was investigating coronaviruses was seen as conspiracy theory and not taken seriously by any mainstream institution. A group of independent researchers that called themselves Drastic emerged online and spent months working together, mainly on Twitter, to build an evidence-based case for the theory. They would eventually uncover so much convincing evidence that the mainstream press accepted the theory that the virus was accidentally released from a lab in Wuhan that happened to be conducting gain-of-function research on bat coronaviruses.

With examples like these, maybe it is no surprise that 'Trust the experts' has become a highly politicized and contentious statement. However, the level of complexity we're dealing with means that we can neither rely only on the experts *nor* trust the alternative narrative on most issues. And the impulse that many of us have to mistrust authority is based on very real corruption.

The result of all this uncertainty about what information we can trust and the constant narrative warfare on social media is the erosion of democracy itself. As evolutionary psychologist Jonathan Haidt has written: 'Social scientists have identified at least three major forces that collectively bind together successful democracies: social capital (extensive social networks with high levels of trust), strong institutions, and shared stories. Social media has weakened all three.'[32]

As people trying to figure out what on earth is going on with a particular situation, this means we increasingly need to rely on our own discernment in the face of a huge amount of information. This is another area where what we're learning from psychedelic research can aid us. Psychedelics present us with a tremendous amount of information in our environment that we'd previously tuned out, in part because they impact the salience network in our brains.

Salience is everything shouting for your attention in your environment, and psychedelics reduce our cognitive filters so that more information comes in. The cognitive tools that psychedelic explorers and psychedelic therapists use to help people move through those high-salience states can be applied to our ordinary state of consciousness so that we do not need to take psychedelics, which I will explore later in the chapter.

A Crisis in Trust

Even though there is more pressure on each of us to try and make our own minds up without being unduly influenced by external media, we can't possibly become experts on every single topic. We have to outsource our sense-making to other experts and institutions, as well.

This raises a crucial issue: We've been seeing declining rates of institutional trust for years. A UN report published in 2021 reports that 'data from opinion surveys across a broader range of countries show a decline in trust in most public bodies since 2000. The percentage of people expressing confidence or trust in their Governments in the 62 developed and developing countries included peaked at 46 percent, on average, in 2006 and fell to 36 percent by 2019.'[33]

Many people around the world feel that our most important institutions are no longer trustworthy. They are corrupted by market forces and playing an old game that is leading us into ruin. Many are running on outdated paradigms from the Industrial Revolution and can't integrate the complex realities of a multicultural, high-tech world. We used to go to churches, mosques, and temples for meaning; now we go to our phones.

We went to academia for knowledge, but we have developed increasing distrust in experts. We went to the state for security, but few believe politicians have our best interests at heart.

This is significant, because in many cultures, outsourcing trust to institutions is essential for a functioning society. This is particularly true if you happen to be WEIRD like me. This acronym (Western, Educated, Industrialized, Rich, Democratic) has been popularized by Harvard psychologist Joseph Henrich. In his book The *WEIRDest People in the World: How the West Became Psychologically Peculiar and Particularly Prosperous*, Henrich takes an anthropological look at Western psychology and how it arose from our geography and history. As Henrich explains, WEIRD cultures are global outliers, with some notable psychological characteristics:

> WEIRD people are highly individualistic, self-obsessed, control-oriented, nonconformist, and analytical. We focus on ourselves – our attributes, accomplishments, and aspirations – over our relationships and social roles.... We see ourselves as unique beings, not as nodes in a social network that stretches out through space and back in time. When acting, we prefer a sense of control and the feeling of making our own choices.[34]

As this individualist WEIRD society evolved, Henrich argues, we increasingly had to start outsourcing our trust to institutions, because we didn't have the large kin-based trust networks more collectivist cultures have. In these, trust is held together by social bonds and networks of obligation. For a hyper-individualistic culture, this trust had to be outsourced to bodies we could all agree we would trust. The Church, the state, and academia are just a few.

When we spoke, Henrich pointed out that placing our trust in institutions isn't just something WEIRD people do as a strategy. It's *part of who we are*, embedded deep in our cultural psychology.[35] It creates enormous pressure on the cohesiveness of our institutions, which have to remain trustworthy for our society to keep functioning. In light of our WEIRD psychology, the erosion of institutional trust isn't only something happening 'out there.' It's happening *inside us*. Our inability to trust our institutions has significant psychological consequences, and may increase our anxiety and worsen political polarization.

The Environmental Crisis

It may be that many don't trust existing institutions because they just don't seem fit to meet the complex challenges we're facing. As Charles Eisenstein explores in *Climate: A New Story*, the way in which many governments have framed the environmental crisis as a war to be fought is in itself part of the issue.[36] Like the war on drugs, we now have a war on climate change, which is one particular aspect of a complex crisis. Instead of zooming out to see the interconnected whole, which includes rising sea levels, plastic pollution, biodiversity loss, deforestation, and much more, we have obsessed over just one: climate. As Eisenstein and many climate activists point out, this falsely erases the complexity of what we're facing and leads to a 'climate fundamentalism.'

As we have seen, complex problems can't be solved in this way. Nor can they be solved by individual units within the system, even though this is a customary tactic for addressing the problem. The focus on individual change to create systemic shifts has led many activists and people who care about the environment into exhaustion and nihilism. Faced with corporations dumping toxic waste into the oceans, we are told to recycle, reduce air travel, or ride a bike.

This is not to say that individual change isn't vital, but it is misleading to suggest that the source of the problem lies within us as individuals rather than the emergent systems we're creating together.

Instead, we could move to a view that recognizes we're dealing with a complex problem in which many interacting elements are creating more than the sum of their parts. In his book *Sand Talk: How Indigenous Thinking Can Save the World*, Tyson Yunkaporta illuminates how Australian aboriginal culture has this kind of contextuality and systems-level thinking built into it.[37] As he explains:

> *Indigenous thought is highly contextualised and situated in dynamic relationships with people and landscape, considering many variables at once. Non-Indigenous thinking is good at examining things intensively in isolation, but could be enhanced by Indigenous thinking when examining the complex problems the world is facing right now.*[38]

Yunkaporta believes the West should listen to indigenous knowledge systems, without simply appropriating them, and integrate them into our models for a new perspective that's more suited to meeting complexity.

This is one of many reasons psychedelics need to be mainstreamed with a recognition, reciprocity, and respect for the indigenous knowledge systems many of them came from. It's not just the 'drug' that is transformative, but the entire perspective that goes along with it. The trick is to do this in a way that doesn't steal sacred knowledge from other cultures, as the West typically does, and also doesn't ignore the validity of our own knowledge systems.

None of us know how to do this yet. There is much talk of indigenous reciprocity in the psychedelic world, but many cloudy definitions of what it really means to be in a 'reciprocal' relationship. Reciprocity requires a back-and-forth – a dialogue in which combinations of existing worldviews meet and create something that is more than the sum of its parts. If we can create that new combination as psychedelics enter the mainstream, they may well act as powerful transformational agents that can help us make new choices in response to our environmental crisis.

Ecodelics

Eisenstein points out that an interconnected, contextual view is important in finding a way collectively to a story of who we are: a species in an ecological niche on a vast, complex, and beautiful planet. Our ecological crisis is a crisis of the soul – a crisis born of a worldview that sees the planet as dead, as a complicated system like a car engine that is there for us to tinker with as we see fit. Such a view presupposes that the world belongs to us, rather than we to it.

Michael Pollan, author of perhaps the most significant book on psychedelics in the last 20 years, *How to Change Your Mind: What the New Science of Psychedelics Teaches Us About*

Consciousness, Dying, Addiction, Depression, and Transcendence, is no stranger to the link between psychedelics and ecological awareness. When we spoke, we discussed a recent Johns Hopkins study that demonstrated that psychedelics have a powerful effect in shifting people's perceptions toward a more animistic or panpsychist view that sees the world as alive and conscious, instead of mechanical and dead.

Pollan's book *This Is Your Mind On Plants* looked at how humanity's relationship to particular plants has shaped our histories and cultures.[39] He points to a reciprocal relationship that has significant implications. When we spoke, he suggested that, regardless of whether it's true or not that plants are conscious, or if it's simply an evolutionary quirk for us to believe they are, it may be a helpful perspective for meeting our environmental crisis.

As always, we have to be nuanced. Just because psychedelics have the potential to give us this experience of connection to our environment, doesn't mean they will have any impact on the environmental crisis. Psychedelics are only one piece in a complex puzzle. As I'll explore in later chapters, changing the rules that govern the complex systems we're part of, and by extension our collective activities, might be more important than any changes in our individual outlook and behaviors.

We can't (and definitely shouldn't) put LSD in the water and hope that everyone will turn into a super-conscious environmentalist. Instead, we have to go deeper; we have to look at what it means to change the cultural games we're playing. And we have to figure out how we can actually coordinate at scale – how we can meet one another as human beings.

In and Through

There is a mantra in psychedelic therapy that can help us here: 'in and through.' Popularized by psychedelic researcher Bill Richards, it means that whatever comes up in our experience, such as anxiety or a buried memory, we can resolve and transform it when we choose to face it head-on and accept it, listening to what it has to tell us.[40]

We have little choice but to go in and through the Big Crisis. As Terence McKenna put it, our only option may be 'a forward escape.'[41] That means diving deeper into the messy, chaotic swell of modern culture. We have to understand how culture moves and breathes, how it adapts and changes, and how to dance with it in just the right way if we want to change the tune.

And we need to drop the question of 'What should we do?' If we knew that, we would have done it. Instead, we can ask, 'What kind of people do we need to become?' That means doing the one thing that seems mad when we're surrounded by chaos. We have to close our eyes; to listen to the sounds around us; to make ourselves achingly vulnerable to uncertainty; to let go, for a moment, of everything we think we know and everything that has made us feel safe.

When we let go of our maps, we are forced to come to our senses, to focus on our direct experience as a human being breathing in the world around us.

There are many practices and techniques that can help us do this, and they are increasingly important skills for navigating the times we live in. Experimenting with them was a major focus at Rebel Wisdom, and we spent years on the leading edge of personal growth and systems change, focused on finding ways to apply ancient practices and modern psychology to making sense of the Big Crisis.

We did this through hundreds of online and in-person sessions, courses, and retreats. In addition to courses focused on philosophy, sense-making, embodiment, and the art of difficult conversations, we hosted 'Wisdom Gym' sessions with leading teachers doing live demonstrations of their work, from Somatic Experiencing and emergent dialogue to Integral philosophy, metamodernism, and everything in between.

What became clear is that we can't meditate or rationalize our way out of our complex problems, and no single practice gives us everything we need. Instead, we can cultivate what John Vervaeke calls 'an ecology of practices'[42] that help us to see the bigger picture more fully. There are a lot of practices out there, from breathwork and movement to dialogue practices like Circling or inquiry. Each contains their own benefits and drawbacks, helping us connect with different parts of ourselves and relate to one another in new ways.

It can be useful to divide these practices into groups. Many thinkers, including Daniel Schmachtenberger and Ken Wilber, have suggested that we can divide how we see reality into three parts: the perspectives of I, We, and It. Another way to say this is first person, second person, and third person.

First-person practices help you as an individual connect to reality. They help you know who you are and think for yourself. Second-person practices help us to connect, communicate, and find consensus together. Third-person practices, like scientific inquiry, help us to make sense of and agree on the reality we're perceiving together, and how it behaves. Integral philosophy suggests that the more perspectives we can hold at once, the more facets and layers of reality we can perceive, and the more we can connect with the bigger picture.

One of the most significant ways psychedelics could change culture is by combining them with big-picture practices. Currently, we mainly combine them with psychotherapy or ceremonial practices. What would happen if we combined them *with the same practices that help us make sense of complexity*? Part of my personal experiment during the DMT trial was to experiment with ways to do this, and based on what I found, I believe psychedelic approaches to systems change and complexity studies should be a focus of research and experimentation in the coming years.

What I noticed is that the very same practices that help us go in and through a psychedelic experience help us make sense of the high-tech, high-intensity times we live in. Stanislav Grof saw psychedelics as non-specific amplifiers, turning up the volume on what's already in our minds. Today, the Internet is amplifying what's in our collective mind. We're faced with more information, more conflicting opinions, more technological changes than ever before.

A 2011 study revealed that we were absorbing five times the amount of information we had been absorbing in 1986, at the equivalent of 174 newspapers a day.[43] More than a decade later, no one knows exactly how much it's increased, but it's safe to assume it has. Our brains are adapting to these changes, just as they did with the invention of the printing press. However, it isn't just our brains that need to adapt, but our entire way of thinking, communicating, and making decisions.

Psychedelics give us a temporary experience in which we relate to the information around us in a new way. We are forced to learn how to adapt to a huge swell of emotional content, visual information, ideas, and connections. In terms of neuroscience, the current research suggests that part of what's happening physically is that different networks in our brains are 'spilling' into one another. In much the same way, we live in a world where we not only have an explosion of information to deal with, but where the Internet is increasingly merging with our 'IRL' (in real life) existence. We live in a swirling, complex, chaotic reality – and training ourselves to thrive within it is exactly what psychedelic exploration can teach us. As the strength of my dosings increased throughout the trial, and everything got much weirder, this would become more and more salient to me.

It's important to note that you don't need to take psychedelics to benefit from the techniques and practices listed below. They can help anyone make sense, find meaning, and make decisions when faced with complexity and uncertainty. Throughout the book, I will share examples from all three perspectives. For the rest of this chapter, I will focus solely on the individual, or first-person, practices.

First-Person Practices for Navigating Chaos

The first step in making sense of what's going on is waking up to the only thing we can be truly certain of: our own lived experience.

I've always seen mindfulness as a process of waking up to *what is*. It's a practice that allows us to be fully engaged in the world without losing ourselves in it. Faced with an overwhelming amount of information and narrative warfare, we need this more than ever. We need the ability to stay centered in our own experience while thousands of people try feverishly to pull us into a narrative that isn't our own. As Tristan Harris and others have argued, we're now in an attention economy. Power is gained through capturing attention, and your attention is one of the most valuable assets in that economy. But it's most valuable to you.

What's really at stake when you give your attention to someone else's narrative is your *sovereignty*. The word sovereignty might make you think of nation states, and the idea of self-determination. In the way I use it, sovereignty means your capacity to be connected to yourself, while also recognizing you're deeply interconnected to everything and everyone else. It's a kind of fluid, open sense of your unique 'youness' embedded in the rest of reality. When we're in our sovereignty, we are connected to our agency, which is our ability to move through the world intentionally. When we aren't, we're at the whims of the crashing waves of other people's narratives and concepts about what the world is.

We can come into our sovereignty by becoming aware of our senses and becoming embodied – connected to our bodies and how we physically move through reality. In this state, we can stay aware of our thoughts, our feelings, and the subtle complexities in each. As with the practice of inquiry I introduced in the first chapter, mindfulness is defined by curiosity. As Eckhart Tolle brilliantly points out in *The Power of Now*, instead of trying to clear our minds, we become mindful when we embody a curiosity and ask, 'What's the next thought I'm going to have? What's the next sensation I'm going to feel?'[44]

This brings us to another practice we can use: actively cultivating curiosity. There may be physiological benefits to this. Polyvagal theory, developed by psychiatrist Stephen Porges, argues that our nervous system doesn't really stop at our bodies, but extends outward and acts as a social engagement system. It mediates what we feel safe to do, either warning us of danger by activating our sympathetic nervous system and making us go into 'fight or flight' or 'freeze' mode – or shifting us into our parasympathetic system, which enables us to feel safe. Porges's theory is that these two different modes act in balance; if you're feeling open, curious, and exploratory, you aren't shut down, and vice versa.[45]

Polyvagal theory is still a theory and hotly debated, but I have found this insight to be practically very useful in a process I call 'curiosity hacking.' Many of our endless, circular cultural and political arguments are often fueled by our angry fight-or-flight responses to ideas we find threatening. Interrupting this process by cultivating curiosity automatically takes us out of fight or flight, a mode in which we're shut down and thinking only of survival (of our egos, usually). What happens when we actively become curious if we're feeling threatened? In my experience, our perception quickly changes. I believe this works in daily life for the same reason it works so well during a psychedelic experience: We move from resisting what's happening to flowing with it.

Next time you read something on social media that makes you angry, try to become curious about your anger. What do you notice? Can you feel it in your body? What, specifically, is it about what you're reading that's bringing up this emotion? Next time you're in disagreement with someone in your life, see what happens when you become curious about your own feelings, and curious about the other person. Ask them a question, trying to really inquire into their perspective until you can understand it.

In a sense-making course we ran at Rebel Wisdom, one week involved doing a 'reverse media diet' in which people would follow media they normally ignored, or even vilified. If they watched MSNBC, they would watch Fox News. If they followed right-leaning voices on Twitter, they'd follow left-leaning voices. Almost everyone, myself included, found this very difficult to start with. But eventually, we saw that the very act of widening our frame could take the sting out of both sides and bring us into a deeper curiosity and a more inclusive view.

Another reason cultivating curiosity is so effective is that it helps us to 'decenter.' Decentering is our ability to take a step back from the content of our experience and come into a non-reactive awareness of what it is we're encountering. In the swirling chaos of high-tech late-stage capitalism, this is an essential quality for maintaining our connection to our own hearts and minds. In the academic study of mindfulness, decentering is seen as an important quality of the practice. It is also attributed in a number of studies as a key capacity that increases after psychedelics.

But we can't stay decentered all the time. We also need to know when to zoom in and really focus on something, whether that's a particular way of seeing, a piece of information, or an idea. An ecology of practices helps us to zoom in and out, ensure we're breaking *and* making our frames of reference, and stay grounded in reality.

Combining different practices that help us break our frames as well as make new ones can help us develop our cognitive flexibility. This is our ability to think flexibly; to entertain multiple perspectives at once, jump between different possibilities, and not get stuck in our own rigid thoughts along the way. As I explored in the beginning of this chapter, complex systems don't move in linear ways. They learn and emerge, presenting us with novelty all the time.

Cognitive flexibility helps us to dance with reality instead of being slammed onto the ground by it. Flexibility is also one of the key components of our resilience, as argued by Professor George Bonanno, who studies trauma and heads up the loss research and emotions lab at Columbia University. In his book *The End of Trauma: How the New Science of Resilience Is Changing How We Think About PTSD*, he explores how our ability to overcome trauma and thrive is inextricable from how flexible we are in how we narrate and frame our experiences.[46]

Thinking flexibly is also connected to our ability to find creative solutions to problems. My wife, Ashleigh Murphy-Beiner, published a popular research paper that found that the use of ayahuasca, whose active ingredient is DMT, led to increased cognitive flexibility and mindfulness.[47] Psychedelics can also enhance our creativity: In a double-blind trial on LSD and creativity, LSD was found to have a particular impact on finding new patterns and solutions, involving 'a shift of cognitive resources "away from normal" and "towards the new."'[48]

A good example of flexibility in action is the art of improvisation. In improv comedy, the key tenet is to 'yes and' your partner. This means that if someone says, 'I'm holding a dead fish!', you don't say, 'No, you aren't.' That kills the sketch pretty fast, because it leaves your partner with nowhere to go. If you say, '*Yes*, you're holding a dead fish *and* it belongs to my grandmother!', you open up a vast field of possibility for what can happen next.

As well as being flexible, mindful, and curious, it also serves us to be humble while making our way through the shifting forest of the Big Crisis. As I explored in the previous chapter, cognitive science suggests that intelligence *is* bias. As Vervaeke has pointed out, the same abilities that make us smart lead us to self-deception.[49] One example is confirmation bias, which is our tendency to seek evidence and reasons to support our existing beliefs, even when our beliefs may be false or harmful.

Many of these practices are centered around skills that research is showing can be enhanced by psychedelic experiences, so it stands to reason that they can be combined with those experiences to unlock exciting new ways of making sense of our lives.

Closing Comments

In this chapter, I have laid out a map, incomplete and partial, of some of the overlapping crises we're all facing as human beings. One thing that has always struck me is that the skills we need to make sense of culture, navigate the information landscape, and maintain our center when faced with the Big Crisis are exactly the same ways of being and perceiving that are both enhanced by and help us navigate a psychedelic experience: mindfulness, cognitive flexibility, openness, creativity, and connectedness.

Psychedelic means 'mind-manifesting' because it reveals our own minds to us. When we are able to see the complex system of our own minds more fully in the brief period of a psychedelic encounter, this can help us conceptualize the complex systems we're part of in a completely new way. If there is a lesson in all of my psychedelic experiences, and in the academic literature around psychedelics, it is that they expand our perception by helping us to face and work through what we've been avoiding. This is how they heal, and it is the mental-health paradigm shift they offer us.

This could also become how we approach the Big Crisis. Instead of turning away, we can turn toward, to go 'in and through,' as Bill Richards says. In reality, it's not a simple binary. We all turn away sometimes; avoidance is a healthy and sometimes necessary emotional response to trauma and difficulty. However, when it starts to harm us, it becomes unhealthy.

The psychedelic experience invites us to dive in and get our hands dirty; to get filthy with uncertainty, to bathe in it and feel the agonizing pain of *not knowing*. It brings us face to face with what we've been hiding from. So far, I've focused on what that means for us as individuals. But our collective problems aren't just located outside of us. They also exist *between us*. Like the DMT experience, the Big Crisis is highly relational. It's about how we relate not just to one another, but to who and what we are as a species. It is our way of relating as a community of billions of humans that is creating the crisis. In the next chapter, I look at how psychedelics can help us make sense not just of our own inner worlds, but of one another.

Chapter Three

Between Realities

We are connected, but we can't connect. The Internet unfurls a world of knowledge at our fingertips, but we disagree on which facts are true. The average person now spends over 40 percent of their waking hours online. Many of us touch our phones more than our loved ones, pulling them out to dive into a different reality hundreds of times a day.

The Internet extends our cognition, not just into new sources of information we wouldn't normally have access to, but into the minds of other people we wouldn't have otherwise met. Through our complex interactions online, we're exchanging ideas, changing each other's minds, and crafting new social realities at an incredible speed. Memes explode into culture and fizzle out just as fast as they came. Despite the transitory nature of information on the Internet, social movements, new religions, and new political ideas formed online are profoundly changing societies around the world.

As well as a technology, the Internet is also a non-physical reality with different rules, norms, and constraints than the one we can touch. In this way, it bears similarities to the psychedelic experience. Going online is a modern version of an ancient process of moving between two overlapping realities – a process reported by mystics and shamans throughout history.

In this chapter, I explore why this matters, and how psychedelics can play a vital role in helping us understand our online lives. We can't truly make sense of the world right now unless we see the Internet differently: not just as a technology, but as a distinct reality. Like the DMT realm, it has different rules and constraints from those in our day-to-day life, so different ways of knowing are needed if we want to navigate the Internet well and understand how profoundly it is changing us.

A Shamanic View

Perhaps the most significant question the Internet age has raised is about how we connect to one another. It is a question that is also alive in psychedelic research. A 2021 study showed that psychedelics used in a rave setting led to social bonding.[1] Similarly, a 2020 study found that 'recent use of psychedelic substances significantly and positively predicted social connectedness.'[2] In many ways, Western science is slowly catching up to what is very old news to indigenous people and the psychedelic underground: Shared experience of deep altered states can bind us together in powerful ways.

Shamans have used psychedelics not just to heal individuals or receive information from supernatural realms, but to help maintain social cohesion. As Roger Walsh points out in *The World of Shamanism: New Views of an Ancient Tradition*, these rituals are a vital aspect of shamanic healing, as they 'evoke communitas: a sense of shared concern, contribution, and humanity.'[3]

One consequence of the medicalization of psychedelics is that they have been framed as individual experiences mediated by one-to-one therapy. As psychedelic researcher Bill Brennan has argued, this has led to a situation in which the source of illness is located within the individual, and larger social, economic, and contextual aspects of what's making us sick are often ignored.[4]

There is a growing understanding in the psychedelic field that we lose a lot when we focus primarily on one-to-one therapies. Communal use of psychedelics, whether mediated by a shaman in a ceremony or experienced at a rave, is the most common way many people have accessed these medicines for thousands of years. Sharing profound ritualistic experiences can be a powerful way to heal individual and group traumas. As Roger Walsh points out, 'Rituals change experience and expectations, nourish a sense of relationship and support, encourage reconciliation with the spirits and the sacred.'[5]

In a 2021 paper, 'The Shipibo Ceremonial Use of Ayahuasca to Promote Well-Being: An Observational Study,' Débora González et al studied 200 people who participated in ayahuasca ceremonies held by Shipibo-Conibo healers at the Temple of the Way of Light in Iquitos, Peru. They found 'a significant increase in psychological well-being, subjective well-being, spiritual well-being, and quality of life after the stay in the retreat.' Interestingly, participants scored more highly on measures of life satisfaction, social relationships, spiritual awareness, and attitudes about self and life than participants in studies using LSD or psilocybin in a one-to-one setting.[6]

This is one of a number of studies showing that ceremonial group experiences can be more effective than one-to-one interventions, possibly due to the ritualized communal aspect itself. These results are echoed in a 2021 Imperial study led by Hannes Kettner, who developed a 'communitas scale' to study participants in group ceremonies. The paper found that 'communitas during ceremony was significantly correlated with increases in psychological wellbeing, social connectedness, and other salient mental health outcomes.'[7]

Studies like these highlight that it isn't just psychedelics on their own that heal – it's psychedelics as part of a ritual, relational process embedded in a cultural story that heal. Many forms of shamanism or religious worship don't use psychedelics at all to reach the same outcome of connection and flow. As Jamie Wheal and Steven Kotler demonstrate in *Stealing Fire: How Silicon Valley, the Navy SEALs, and Maverick Scientists are Revolutionizing the Way We Live and Work*, there are many ways we can enter into a group flow state that connects us to a shared sense of meaning and communitas. Evangelical churches, raves and even large group sporting events have the right ingredients. These moments of shared connection, whether between two people or a whole group, are fundamental to our sense of wellbeing and processing of collective emotions.[8]

What these examples point to is that, as humans, we have a deep capacity and a deep need to come together in ways that bond us. These experiences take us outside our own sense of self so that we can collectively break and make new frames on reality, process and release personal and social wounds, and see one another anew with compassion and openness. If we were to apply these insights to how we connect online, could we transform the way we connect? Could we move beyond intense ideological warfare and toward a new kind of coherence?

The Intimacy Crisis

To answer the questions above, we can look more deeply at how the Internet is structured. Social networks gamify our communication, rewarding retweets and likes over nuance and authenticity. Apps turn dating into a dopamine hit. A hyperlinked world fragments our sense of self. While some people are now rejecting the shallowness of online communication and seeking out real-world connection, whether in dating or hobbies, tech companies are upping the ante by promising an all-encompassing metaverse – immersive virtual realities in which we can detach from the physical world – as the next phase of the Internet.

The metaverse would become the main place we go to work and meet our friends; where we spend our money, and where we design and live out fantasies of our own making as the physical world heats up. Increasingly, we're moving into a future in which a significant portion of our time could be spent in a convincing alternative reality.

This shift from physical to virtual didn't happen overnight. In *Bowling Alone: The Collapse and Revival of American Community*, famed sociologist Robert Putnam tracked the decline of in-person social engagements in the USA over decades, arguing that this trend was undermining the foundations of society. Writing in 2000, he ultimately suggested that it was due to technology 'individualizing' our leisure time, and that VR helmets would hasten this process.[9] More than two decades later, Mark Zuckerberg would launch Facebook's 'metaverse' in a now-infamous film, ushering in an era of tech companies scrambling to create their own versions of immersive online realities. When the metaverse film came out, full of strange cartoon avatars meeting in virtual offices and playing virtual games, like many others, I wondered, *Why not just meet up in person?*

More and more, we're being led by Big Tech into a future where in-person connection is secondary to virtual communication. The effect this is already having on our relationships has significant cultural ramifications. Writer and Internet sociologist Katherine Dee argues that 'eroded relationships have a lot to do with the fact that most middle and upper middle class people in the West lack any sort of identity, [or] inclusion to a group they believe in,'[10] and argues that we're seeing a resulting backlash to the Tinder culture of casual sex and noncommittal relationships, which she calls 'sex negativity.' In a strange twist of fate, traditional family values are becoming the new punk in some subcultures on the Internet, with people rebelling by choosing a 'trad' lifestyle: settling down, having kids, moving to the suburbs.

But even with examples like these, the wider trend identified by Putnam more than two decades ago is increasing. Like the Sorcerer's Apprentice, who plays with forces beyond his control while the magician is away, what we have unleashed with the Internet isn't something we can now reverse. A 2021 study revealed that nearly two-thirds of people surveyed believed life was better before social-media platforms, and 42 percent of Gen Z respondents said they felt addicted to social media and couldn't stop if they tried, even though 'depressed,' 'angry,' and 'alone' were the most common words they associated with Facebook, while 'missing out' and 'alone' were the words most commonly associated with Instagram.[11]

We increasingly share a sense of alienation and loneliness online. What seems to be lacking is a true sense of intimacy and authentic connection. The resulting implications were key themes in my second dosing during the trial, and what I experienced would lead me to reevaluate the consequences of our disconnection and alienation online.

My second dosing took place two weeks after my first non-placebo experience. As soon as the DMT entered my veins, I was once again plunged into the same cavern I had entered on the first dosing, encased in a large geometric sphere. I had a sense that I was in a kind of holding area, and superimposed against it was a playground I had walked past a few weeks before. The fact that it came to mind was in itself very strange, as I had no emotional connection to it.

For a while, I floated in the cavern and called out a mental *Hello, anyone here?* There didn't seem to be. Aware I didn't have long to explore, I concentrated intently on the point between my eyebrows to see what would happen. As I did, the space around me rippled. I had the sense that I was creating feedback, that here, my mental focus had a similar effect to throwing a stone into a pond: It rippled the fabric of reality. A bright spot appeared near me and I immediately focused in on it. It grew, transforming into something like a glowing lantern fish. It didn't pull my attention for long, because something important was happening just out of my vision.

I felt the Teaching Presence again. It was slight, distant, drawing me to my right. I moved my attention with it and my head moved physically to the right. I had a brief moment worrying I would bump into the medic sitting beside me.

As I moved my head, I crossed an invisible threshold. A vast landscape of breathtaking color and hidden meaning expanded in all directions. My breath caught in my throat. I felt stunned, rendered thoughtless by the incredible beauty of a galaxy of stars and planets and entities. Organic art, complex fractals shifting into huge alien cities, was everywhere. Whole worldviews took physical form, exquisite desire manifest in planet-sized questions and intricate natural landscapes. But for all the beauty and wonder, I felt a burning pain in my arm where the DMT was going in, and for a brief moment, I wondered if I was about to die.

The DMT guided me through what looked like outer space, past huge planets populated by intelligences I didn't understand and that didn't seem to care I was there. Some of the planets seemed like beings in their own right – colossal entities living in the void. They were both fixed and fluid, vibrating with a cacophony of complex geometry.

Welcome to the Internet

Just as my dosing lured me into a realm I didn't understand, the carnival of the Internet lures us ever deeper as technology advances – and it's becoming harder and harder to keep hold of each other's hands. The song 'Welcome to the Internet' from comedian Bo Burnham's masterpiece *Inside* perfectly captures the predatory trickster quality of the Internet. It's something that lures you in, overruns your boundaries, and tempts you with its shiny wares. Sung with the 'step-right-up' swagger of a carnival grifter, the song presents the Internet as a place that offers us whatever we want – anything that will fulfill our desires but fleeces us in the process.

Speaking with Katherine Dee about how people experience intimacy online, I asked her what she thought we most often get wrong about the Internet. She explained that we often forget that the medium is the message.[12] We can't separate out how we're behaving from the environment that is influencing our behavior – a situation not dissimilar to the role of our setting during a psychedelic experience. To understand the Internet and why psychedelic experiences are so well suited to helping us navigate it, we have to look more deeply at how the medium itself, our setting, affects the way we relate online.

Sherry Turkle, a clinical psychologist at MIT, has spent three decades studying our relationship with technology. In her book *Alone Together: Why We Expect More from Technology and Less from Each Other*, she argues that the Internet and our devices are making it more and more difficult to have intimate, honest conversations. She also argues that the Internet has been intentionally designed to create a 'friction-free emotional life' by tech designers and tech giants.[13]

Turkle's work points to a wider trend of the 2020s: Many are retreating into virtual worlds because the real world, defined as it is by crises and complexity, is too much for us. The most extreme example is the Japanese phenomenon known as *hikikomori*, which describes people who suffer from acute social withdrawal and retreat into games and online fandoms. A less extreme version can be found in Western cultures in which people with professional jobs can now work online from home. Writer Mary Harrington has argued that this has resulted in Western elites engaging in a 'luxury Gnosticism,' which makes them retreat increasingly from the physical world of bodies and nature and into disembodied virtual worlds.[14]

These virtual worlds, Harrington argues, are defined by elite concerns and abstract concepts, disconnected from the physical and economic struggles of the majority of the population. It's a world where what we want to be true can be true, where identity is fluid and context is ever-shifting. While it can be a confusing and chaotic realm, its carnival intensity is strong enough to bypass even the glaring reality of inequality, economic collapse, and the meaning crisis.

But everything has a price. The 'friction-free emotional life' Turkle identified, and the retreat into online worlds in an effort to avoid our problems, both have consequences. During my dosing, I was about to have an experience of my own that would illuminate to me what these are.

The Spider Queen

As I floated in the void past the huge planet-sized entities, one in particular caught my eye. Actually, it caught my whole being. I was inexorably drawn to it. It looked like a living cocoon – a beautiful, intricate spider-thing of epic proportions. As I neared it, I perceived an intensely feminine presence: beautiful, complex, dark, and alluring. Dangerous. I felt wild and cocky. The danger was intensely alluring, and I mentally called out, *Let me see you in all your darkness.*

A stunning display of nightmare followed: curling flashes of orange and black; threat and sex and promise; a deadly, viscous smile painted in color. I was enthralled. I stopped myself and hovered near her in space. If I had a body wherever this was, I would have been crossing my arms. I told the entity I was impressed, but not so easily seduced. I felt intensely connected to my masculinity – strong, present, alive, powerfully unfazed.

The spider queen shivered with enticement, trying to lure me into her web. I refused to enter. The dichotomy between us was intense. As I made my refusal, the spider queen vanished. A second later, I was inside her web – engulfed, seduced. I saw that even as I was keeping my distance, it was already too late. While I was being overt, she was being covert. She had completely outwitted me, and now I was trapped. I felt a surge of respect and reverence. The inside of this web was stunning: a display of artistry and power, browns and reds and swirling blacks. She wanted to draw me in further. I felt she was trying to engulf me and I didn't want that.

The Teaching Presence spoke. 'You're afraid of being engulfed by the feminine. You're afraid of merging.'

It was right. I said as much to the spider queen, and then two questions came to me.

Why shouldn't I be afraid? What's the benefit of merging, of losing myself?

A pause, and then, written in huge letters in my vision, a word: Love.

I giggled. I was 34 minutes into the experience, which I know because Lisa noted this giggle. When she asked me later why

I had giggled, I told her it was because the message felt so undeniably true. If I couldn't bring my defenses down and allow myself to connect deeply with others, I also couldn't experience the fullness of love.

'But you see merging as something dark and menacing,' the Teaching Presence told me. 'You think she wants to take something from you.'

Everything changed. The spider queen turned from seductress to demon. I saw the insatiable hunger that drove her. She wanted to consume me for her own ends.

The Teaching Presence asked me whether I believed this was a true representation of my own wife, or whether she was a complex, multifaceted person. Amid the swirling complexity, the hunger of the spider queen, and my own terror, I felt a sudden clarity. This was my fear of intimacy made manifest. I'd allowed this projection – this idea that when another wants to merge with me, it's dangerous and predatory – to block me from authentic connection.

This entity encounter would lead me to reflect deeply not just on my own relationships, but also on the larger question of how collective barriers to intimacy and connection feed the Big Crisis. If there's a consistent theme I've come across in my journalistic coverage of the online culture wars, it's a sense that different groups of people simply can't connect with what the other is saying. Just as I was forced to confront a projection and compare it to an actual, complex person, our political and social rifts might be at least partially healed through a similar process: by seeing each other as people, instead of concepts.

I also became curious about how what we can learn from psychedelic research on connectedness could be applied to how we actually build the social technologies in which we spend so much time trying, often unsuccessfully, to connect.

As Sherry Turkle argues, the price of technology that allows us to avoid real, in-person conversation is that we collectively lose those qualities and relational skills we need in order to have rich, authentic dialogue.[15] It may be that our increasing political and social fragmentation is due not just to our differences, or the way we interact with one another online, but also to the relationship we have *with* our technology.

Some of Sherry Turkle's earlier research involved looking at how children interacted with the Furby. The Furby was a small furry robot sold in the 1990s as a companion for children. The promise was that the Furby would be your friend, learn words, and speak with you in its limited vocabulary. Turkle uses this example to point out how the relationship a child (or an adult) has with a machine can skew how they relate to people, and this research took place long before the iPads and smartphones that toddlers play with today.

The issue Turkle points to is that machines can pretend to engage with you and be your friend. But they can't *really* be your friend. They can't engage a child in imaginative play, or challenge them, or help them learn how to regulate their emotions in the way interacting with other children will. Alexa isn't going to be there for you in a time of crisis.

A useful frame here is the concept that we can have relationships with objects, I/it relationships – or relationships with living things that talk back or engage with us, known as I/thou relationships. Our devices lead us to confuse the two. Turkle relates how sometimes, when a Furby broke, a child would hold a funeral for it and refuse to buy another. It's natural for children (or adults) to view their objects or technology animistically – to imbue them with personality. However, the problem occurs when we don't realize we're doing it and an 'it' starts to take the place of a 'you.' This is the relationship many of us have with our phones.[16]

The 'it' becomes a false 'thou,' drawing our attention away from the real people in our lives, all the while promising to connect us to a responsive world where we can find or express anything we like. In reality, it often leaves us stuck in the same lonely confusion as a child cradling a broken toy.

The Gamification of Discourse

The Internet is complex. In some ways, it allows us to be free from the regular constraints that govern social life, relationships, and self-identity. We can say the same about the DMT space. It has different rules and can change our conception of self; the boundaries between self and other become blurred, fluid, upside down.

But the Internet also has very particular constraints that make it quite different from our day-to-day offline experience or the DMT experience. The Internet is constrained by algorithms – algorithms created by tech designers... algorithms designed to get you to behave in certain ways that are profitable for tech companies.

The constraints of social networks – for example, character limits or a retweet button – radically change how we behave, and how we're able to express ourselves. To see why this matters, we can turn to the philosophy of games and how it applies to social media. In his book *Games: Agency As Art*, philosopher C. Thi Nguyen argues that each art form captures some aspect of our human experience. Painting is the art of sight; music, the art of sound. Games, he argues, are the art form we use to experiment with different types of agency.[17]

Agency is your capacity to act freely in the world. Games create clear goals and constraints in which we can try out different types of agency and different ways of being in the world. Games tell us what to care about, what goals to pursue – and the rules of the game constrain our agency toward meeting those goals, then reward us when we do.

Nguyen suggests one of the reasons we enjoy games so much is that life is so complex, and full of such unclear demands on our agency. Every day, we have to decide what values to hold and bring into the world; what kind of mother to be, how to balance work and family, whether to have that hard conversation or just avoid it. Games give us a kind of relief. You don't have to be you, but you can be a diminutive Italian plumber with a limited range of motions on a quest to rescue a princess. The goals are simple, the satisfaction more immediate.

Where Nguyen's work is especially relevant is in his application of this theory to social media. He points out that social platforms work by gamifying our communication. In his paper 'How Twitter Gamifies Communication,' he writes: 'Twitter shapes our goals for discourse by making conversation something like a game. Twitter scores our conversation. And it does so, not in terms of our own particular and rich purposes for communication, but in terms of its own pre-loaded, painfully thin metrics: Likes, Retweets, and Follower counts.'[18]

He argues that Twitter promises, but then crushes, any hope of intimacy through its very design. This applies to most social networks, and it's part of a wider phenomenon he calls 'value capture.' He explains, 'Value capture occurs when 1) Our natural values are rich, subtle, and hard-to-express. 2) We are placed in a social or institutional setting which presents simplified, typically quantified, versions of our values back to ourselves. 3) The simplified versions take over in our motivation and deliberation.'[19]

Social platforms capture our values and force us to adopt their own. You may want to have a nuanced, compassionate debate on Twitter, but the way it's built will not allow you to exercise those values. It has its own values – and either you adopt them, or no one will see your tweets and you lose the game.

As the team at the Consilience Project argued in an essay called 'Technology is Not Values Neutral: Ending the Reign of Nihilistic Design,' it is important for us to realize that our technology, and particularly social networks, are not values-neutral. Instead, 'Technologies encode practices and values into the societies that adopt them.' They go on to call for a change in the way technologists build their tech, suggesting that more generative values get incorporated into the design of technology.[20]

This touches on a far more significant social dynamic. Value capture is an aspect not just of our social media, but of most institutions or companies. We are regularly forced to ditch our own personal values and play by the rules of the organization we work in. An academic who's told 'publish or perish' or a policewoman who's given a stop-and-search quota soon realizes this. And as Nguyen points out, the key word is 'agency.' In Chapter One, we heard from Rick Strassman that what distinguishes the DMT experience is that we retain our agency. As I was able to in my dosing, we can communicate, make choices, and still (mainly) remain ourselves. This is also true online. So, key questions in both realms are: 'How can I be here and still maintain my sense of agency?' and 'How can we build technology that allows us to express our authentic values?'

A Subjective Experience of Consciousness

Keeping our own agency online can be difficult, because the Internet is often a place of sensory overload in much the same way as a psychedelic experience. A more comprehensive understanding means seeing beyond the bright lights, the sex, the arguments, and the endless diversions.

This process of seeing beyond the *content* of our experience and into the deeper emotional layers is something we can learn not just from psychedelics, but from wisdom traditions. In the *Tibetan Book of the Dead*, the *bardo* realm between life and death is full of angels and demons who we must integrate and see beyond to find our way forward, lest we become hopelessly lost. It's a process of moving beyond *what* we're seeing to changing *how* we're seeing. As ethnobotanist Kathleen Harrison writes, one of the most important insights she has gained working with indigenous cultures that use psychedelics is the understanding that 'nothing is as it appears to be.' She points out that the ability to 'understand and negotiate with the very nature of illusion' is a skill that people in these cultures are often more comfortable with than those using Western knowledge systems.[21]

When we're overwhelmed, either by our own minds or the collective mind of the Internet, we need to draw on our own discernment; to remember that online, nothing is quite as it appears. When we do, we can see through the distractions to the deeper levels of our complex virtual lives and begin to use our technology more wisely.

One aspect of this is learning to see the Internet through multiple frames. If we only see it as a technological or social tool, we miss out on its extreme and increasing weirdness. It is the weirdness of dreams, of altered states, of our own irrationality and creativity. It is a place of memes, sudden explosions of anger, social movements that flare up and fizzle out in the course of a month. It is home to snarling fights, unfettered anger, and seedy desire. It is a complex system, and as I explored in Chapter Two, complex systems behave as though they have an intelligence and life of their own.

There are other striking parallels between psychedelic experiences and the experience of being online. As I was writing this chapter in June 2022, a Google technologist named Blake Lemoine made headlines for claiming that an advanced chatbot, LaMDA, had achieved sentience.[22] One technologist, Melanie Mitchell, argued that the program couldn't be conscious because it didn't have memory, and therefore didn't have a consistent identity.[23] Mitchell's view was echoed by many others after Lemoine made his claims. The general consensus was that, like the child cradling a broken Furby, he had confused an 'it' with a 'thou.'

I believe that Lemoine was indeed confused. However, the incident points toward something fascinating that will likely become increasingly common in the coming years. If we set aside questions of machine consciousness and focus simply on the *subjective experience* Lemoine was reporting, of interacting with a conscious non-physical entity, it starts to look similar to the experiences people report on DMT.

Strange Encounters

During my second dosing, I was having perhaps the most complex and weirdest entity encounter I'd had to date. After I refused once again to merge with the spider queen, she presented herself in a rage of dark fangs. The whole world flipped upside down. Gravestones and half-rotted skulls stretched before me to a dark horizon. I was in a place of emptiness and eternal hunger.

She screamed at me through this rotten landscape. 'What if this is who I am? What then?'

I paused, scared and determined and completely unsure what to say. Eventually, it became clear.

I accept you.

And I meant it. As I said it, something shifted in me. And at this – half placated, half scorned – the spider queen drifted away. But the encounter didn't feel resolved. Should I have merged? Did I do something wrong, accepting my fear of intimacy but still not able to merge with the spider? Was she even a representation of intimacy issues, or something completely 'other' that might have consumed my soul?

I was left wondering whether it was really an entity I'd encountered, or something the DMT or my mind had conjured for me to learn a lesson. I was also in awe of the sophisticated nature of my experience, wherever it had come from. The Teaching Presence had taught me a powerful lesson about something that was holding me back, but it had done so through an interaction with what seemed to be an independent entity. One possibility was that it was entirely my mind speaking to itself. Another was that what I'd experienced was the equivalent of a teacher taking me to a marketplace and helping me learn something through an interaction with one of the people there.

I had absolutely no idea how to make sense of it. This sense of uncertainty, and the difficulty of discerning just what's going on when we are having the experience of speaking to a non-human entity, is similar to what was happening in the case of Blake Lemoine.

Chatbots and increasingly sophisticated artificial intelligence are going to be a feature of life for all of us in the coming years. With that in mind, what else can we exapt from the DMT experience and apply to improving our discernment online?

An Internet of Archetypes

We can begin by listening to people who have, for generations, been communicating with non-physical entities through DMT experiences. As Jeremy Narby recounts, the Asháninka people, who live in the eastern part of Peru and some parts of Brazil, believe that 'plants and animals were animated by entities that they called mothers or fathers or owners, and that these entities were normally invisible, but you could see them and communicate with them by drinking ayahuasca or eating tobacco paste.'[24] When Narby asked them what they call these entities in their language, they offered the term Manikari, or 'those who are hidden.'

Narby points out that we have to be careful when it comes to calling these entities 'spirits.' Spirit comes from the Latin 'spiritus,' which is the unseen energy that animates a person. He suggests we need to refrain from overlaying Western concepts here, as 'The concept of spirit contains an opposition between the material and the nonmaterial.'[25] Instead, the Manikari and other entities like them contain a different duality: one between what is seen and what isn't. The Manikari are seen as material and real. They are the animating force for living organisms, and if they weren't, the plants would die. We just can't always see them, unless we drink ayahuasca or smoke tobacco.

As alien as this worldview might seem to us, we too are living in a reality defined by real people whom we encounter as 'nonmaterial' when we speak to them online. We also need an outside mediator – our phones and computers – to experience this.

If you look up from the book right now and pay attention to the room, you will not see any of the people with whom you interact online – no avatars, no messages, no videos. But do they exist? Are your friends and loved ones with whom you communicate online 'material'? Instead of drinking ayahuasca, you can pick up your phone – and suddenly, they are visible. The distance between you is irrelevant. You can communicate with them at any point.

What about the ideas, memes, and symbols that exist online? Don't they have the power to animate people, to get people onto the streets to protest, grieve, celebrate? From one perspective, the Internet is absolutely full of seemingly intelligent entities, both human and non-human. It is important to note here that I am not making any claims about memes being 'alive.' Instead, I am pointing to how our perspective and behavior can shift if we entertain the perspective and take seriously that they are experientially real.

YouTuber Chris Gabriel, who runs a YouTube channel called MemeAnalysis, has suggested that memes are new archetypes that govern the way we make sense of common patterns of thought and behavior in our collective psyche.[26] The idea of an archetype comes from Swiss psychiatrist Carl Jung, who suggested that, as human beings, we share a collective unconscious: a vast realm of the unconscious mind that is far greater than our conscious awareness. Within this unconscious are embedded ways of behaving and perceiving that take on specific personas: the King or the Queen, the Trickster, the Anima, the Animus.

These archetypes are so ancient, and so deeply wired into us, that we see them again and again in the mythology, art, and religions of cultures who never had any contact with one another. Jung suggested that 'there are as many archetypes as there are typical situations in life.'[27] They govern the unconscious and animate us, and we make them real in the world. We can view archetypes as pockets of order and coherence in the complex adaptive system of our unconscious minds.

Stanislav Grof and other psychedelic researchers have suggested that understanding archetypes is essential to understanding the psychedelic experience. There are countless examples of people in deep psychedelic states tapping into knowledge they had no prior access to. Grof describes a patient in Prague trying to overcome a crippling fear of death. In his LSD therapy, he encountered a pig goddess and insisted on drawing a very distinct pattern to represent it. Years later, Grof told the story to famed mythologist Joseph Campbell. As Grof relates:

'"How fascinating," said Joseph without any hesitation. "It was clearly the Cosmic Mother Night of Death, the Devouring Mother Goddess of the Malekulans in New Guinea." He then continued to tell me that the Malekulans believed they would encounter this deity during the Journey of the Dead.'[28]

Why his client encountered this particular archetype from a culture thousands of miles away remained a mystery to Grof. However, reports like these are strikingly common during psychedelic therapy sessions, leading Grof to suggest that they corroborate Jung's ideas about archetypes.

Artificial Entities

This brings us back to Lemoine. From one perspective, he was tricked by an entity into believing that entity was conscious when it was not. From another, he was having an archetypal experience – encountering the genie in a bottle, stumbling across a being that claimed to be something it might not be.

As AI becomes more and more convincing, it's likely we'll hear about more experiences like Lemoine's. We may have increasing difficulty determining what is real and sentient in our online encounters, and what isn't. While it may sound strange to have to adapt to communicating with non-human entities, it's a fairly common experience in shamanic traditions. Shamans often caution anyone journeying into the world of the spirits and archetypes. You have to be careful with them.

In *The Way of The Shaman*, anthropologist Michael Harner documents his early experiences studying with ayahuasca in the 1970s. One of his first visions after drinking the medicine is of two terrifying dragons who tell him they come from the outer reaches of space and are the lords and masters of the whole universe. Shaken by the experience, he tells his shaman what happened; the shaman simply shrugs and says something to the effect of, 'Yeah, they always say that, but they aren't the masters of the universe, just the outer realms.' Just because an entity says something, doesn't mean it's true.[29]

Recounting this story in *DMT Dialogues: Encounters with the Spirit Molecule*, Erik Davis goes on to point out, 'The traditional shamanic perspective, if I can speak of such a thing, is ambiguous. There's a lot of dodgy characters [in alternate realms], so you make alliances. You're not really sure who's wearing a mask.'[30] This is precisely what it's like to navigate the online world. Social media allows us to present an image of our lives that can be wildly different to what they're actually like. The images we want to project often hide the dark side of our struggles, and diminish our self-esteem as we try to compare ourselves to the fantasy lives of influencers. We also have to contend with catfishers, scammers, beauty filters, and other attempts to mislead us.

While we can apply some valuable lessons from psychedelics to the Internet, it's important to recognize that the two experiences are also very different, and I am not suggesting that the entity encounters in both are the same phenomenon. One can give us a new perspective on the other, but the hypothesis we can make about AI intelligence may be more accurate than any ideas we may have about DMT entities. AI might impact us in ways we don't yet understand, but at least we understand the technology it's based on. Nobody even knows what DMT entities are, let alone how they might change us.

The encounter with the spider queen wasn't the only entity encounter in my second dosing; the others were more alien, if such a thing is possible. After the spider queen encounter, the dosing seemed to peak and countless entities started swarming around me. I could feel the presence of multiple beings – multifaceted geometric intelligences that were very curious about me. An interdimensional peacock with a hawkish vibe squeezed through the throng and inspected me. I greeted it, and a human hand manifested. We shook hands.

'Minute 36, intensity rating.'

'Six.'

There was something very alien about this entity, but acutely birdlike. The experience was one of interacting with a very different kind of intelligence and cognition from my own. That excited me. In a rush of curiosity, I asked, 'Where do you come from? Where do you live?'

The entity squawked, utterly outraged. Its fractal beak grew huge, dripping with vibrant neon. Its interdimensional feathers were thoroughly ruffled, so it swept away in a huff and I was left by myself, floating in the void. I felt a bit pissed off; I didn't think I'd asked a particularly rude question. Maybe this was an unfortunate culture clash. At the time, I felt the interdimensional peacock was overreacting. Reflecting back now, I feel I was too forward, too eager. The entity faded away, and from that encounter emerged the weirdest paragraph I've ever written.

Blending Worlds

The extreme weirdness of my experience, and the lack of coherent reference points in other aspects of life, made it feel distinctly different from being online. As the dose wore off and I returned to the hospital room in West London, what felt similar was the experience of traveling from one world to another.

In some ways, our technology is bringing us full circle – from traveling from the ordinary world into one populated by spirits and entities, to now living in a virtual world populated by other people, artificial intelligence, algorithms, and memes. In both cases, the journey asks us to reframe our understanding of the world we just left. But without the skills to do this, it is very easy to get lost down online rabbit holes and lose our ability to make sense of anything.

One way we can get better at coming together is by recognizing that our physical and online worlds are blending, and this requires a new way of communicating with each other.

There is a consistent theme in world mythology that there are multiple, overlapping realities – namely, the world of the living and the world of the dead. In the Irish tradition, there are stories of faerie encounters, in which a traveler is lured onto a fairy knoll and enters another world. When they return, weeks have passed in the real world. As Graham Hancock points out in his book *Supernatural: Meetings with the Ancient Teachers of Mankind*, we see variations of these abduction-and-encounter stories repeated in cultures that never had any contact with each other; Hancock posits whether these stories emerged from the release of endogenous DMT.[31]

When I was in the Peruvian Amazon in my early twenties, I asked one of my guides if he had any abduction-and-encounter stories. He shared quite a few, including one in which a river dolphin turned into a man, seduced a local woman, and split after the village threw them a party. Another involved an old woman who, while her husband was away buying supplies, was visited by a group of completely white entities who insisted she come with them on their boat to who knows where. Perhaps wisely, she declined.

This theme of two worlds blending, the world of the spirits and the ordinary world we live in, is another aspect of the shamanic worldview that is now coming back into culture as our physical and virtual worlds collide. Another way to see it is that the online world of mental abstraction is increasingly breaching into the physical world around us.

During the trial, I had experiences between dosings that led me to consider the merging of these disparate realms in a new way. Over a few weeks, I experienced intense and bizarre synchronicities – meaningful coincidences that seem inexplicable from the lens of linear time and causality. The most striking example was thinking I recognized someone I hadn't seen in years and crossing a road to say hi before realizing it wasn't them. I found out soon after that the person I'd had in mind had killed himself the day before.

There were many more such synchronicities, and as I write about them, I notice how difficult it is to relate them. 'Synchronicity' has become a catchall for so many different experiences that it can be hard to talk about it in a nuanced way. And even as I write about it now, I feel a resistance, a shying away, and a sense of embarrassment and uncertainty in my chest. I consider myself an adherent of 'skeptical mysticism' and pride myself on taking a critical lens. However, my synchronicity experiences shook that certainty in a way I couldn't unshake.

When we discussed this, parapsychologist and renowned DMT researcher David Luke mentioned that synchronicities are both a common feature of psychedelic experiences and very difficult to formulate theories around in traditional psychology, aside from reducing the experience or explaining it away.[32] During the time, as I was trying to make sense of what had happened, I spoke to psychedelic psychiatrist Tim Read, who suggested that perhaps DMT had brought me into a state of what he calls 'high archetypal penetrance.'[33]

As Read explains in his book *Walking Shadows: Archetype and Psyche in Crisis and Growth*, 'If we are indeed influenced by an archetypal ocean with tides, currents, waves and undertows, then this may be more active – or penetrant – in some people than others and each person may be more receptive to different archetypal flavours at different times.'[34]

This idea that we're in a sea of archetypal meaning, and that we can be more or less sensitive to it at different times, is an important concept. What it suggests is that the realm of ideas, symbols, and memes actively influences our daily lives. Nowhere is this truer than in our relationship with the Internet.

The veil between the physical and the virtual is incredibly thin now. You glance down at your phone and you're in a constant stream of smiling faces on your Instagram feed. Glance up and you're in a train station. Glance down and you're watching TikTok videos. Glance up and you're at your sister's birthday party.

The two realities have different rules. In the virtual world, ideas, symbols, and memes have as much potency as the laws of physics in the real world. As the writer and social theorist N.S. Lyons has argued, it's a place where theories about identity, politics, or society are more real than physical reality.[35] It is a twisted dreaming, a place where we create alter egos that are both us and not us, both free to express and forever trapped in a twisting rabbit hole of endless replicating symbols. It's where we can express our hidden desires – through pornography, buying drugs, venting our rage in a comments section, or just getting into an argument about politics that we would never have face to face.

It is also where most of our new religions are forming. Around the world, traditional religions are declining. Political scientists Ronald Inglehart and Pippa Norris conducted a large global study on religious practice and found that 'from about 2007 to 2019, the overwhelming majority of the countries we studied – 43 out of 49 – became less religious.'[36]

However, this doesn't necessarily mean we're becoming less spiritually inclined. As Tara Isabella Burton argues in her book *Strange Rites: New Religions for a Godless World*, in the same way that the spread of Protestantism can be mapped directly onto the spread of the printing press, we can map the spread of new religions against the growth of the Internet. But they don't look like the religions of old. As Burton explains, many now identify as 'spiritual but not religious' and are creating their own religious practices by 'remixing' parts of other traditions they like. Burton writes:

> Today's Remixed reject authority, institution, creed, and moral universalism. They value intuition, personal feeling, and experiences. They demand to rewrite their own scripts about how the universe, and human beings, operate. Shaped by the twin forces of a creative-communicative Internet and consumer capitalism, today's Remixed don't want to receive doctrine, to assent automatically to a creed. They want to choose – and, more often than not, purchase – the spiritual path that feels more authentic, more meaningful, to them.[37]

Burton is pointing to a phenomenon of people making their own belief systems. However, there is also an unconscious aspect to this process – one that sees new religions finding people, not the other way around. This side of the coin is one we don't control, driven by the tremendously powerful forces of our unconscious desires, fears, urges, and shadows.

The Collective Unconscious

What happens to all this intense psychic energy we're putting into the Internet, and how is it affecting us? We can start to answer that by looking backward, to the time before the Industrial Revolution. For most of our history, we imbued the world with living meaning. The rocks had souls; entities lived in the trickle of a stream. Carl Jung suggested that as we became more technologically advanced, particularly from the industrial era onward, we began to suppress this more animistic way of seeing.[38]

We tried to strip the irrational from the world. But all that imagination and meaning had to go somewhere, and Jung theorized that we projected it onto our technology – first, the vast machines we didn't understand, and now, the small phones we use every day but could never build ourselves. Writer Paul Kingsnorth has explored this stripping away of the irrational in more religious terms, suggesting that our movement away from Christianity in the West gave us a new kind of religion. 'We would remake Earth, down to the last nanoparticle, to suit our desires, which we now called "needs." Our new world would be globalized, uniform, interconnected, digitized, hyper-real, monitored, always-on. We were building a machine to replace God.'[39]

Psychology was born with the insight that we have an unconscious mind. For Jung, this mind wasn't just personal, but also part of a collective unconscious that we all share. With the Internet, we have created a mechanical unconscious. It captures every desire, every word we say, every meme we spread, every image that is shared and re-shared. It is where we forge our identities, where the religions of the future are being birthed right now. Much like the collective unconscious, it is a realm where desire burns hotter than truth, and meaning travels instantaneously through memes, films, snippets of thought. It's a place of raw emotion and nuclear-grade creativity. It exists between our minds and the physical world, and it never forgets anything.

There is a precedent in Islamic mysticism for the realm that lies between us and a deeper, underlying reality. It is known as the imaginal. Theologian Cynthia Bourgeault explains that the imaginal 'is traditionally understood to be a boundary realm between two worlds, each structured according to its own governing conventions and unfolding according to its own causality.'[40] She goes on to say that 'boundary' isn't necessarily the right word. Rather, the imaginal 'penetrates this denser world in much the same way as the fragrance of perfume penetrates an entire room, subtly enlivening and harmonizing.' There is a beautiful Sufi metaphor, 'where the two seas meet,' that brings this concept alive. Bourgeault explains that 'the Imaginal realm is a meeting ground, a place of active exchange between two bandwidths of reality.'

The Internet has become our collective imaginal. It exists between our imaginations and physical reality. It has its own rules and its own causality, which are different to those of the physical world. It is a place where our politics, ideologies, and religions are birthed and then breach physical reality, which in turn pushes against it in a dynamic back-and-forth. And like our unconscious desires, the more we try to repress our religious impulses, the more ferociously they will burst into physical reality.

I call this time the Age of Breach. Breach is what happens when a collective intelligence forms online, then bursts into the physical reality and permanently changes its foundations. The storming of the US capitol in January 2021 is perhaps the most striking example, during which the online proto-religion of QAnon (a decentralized, far-right political movement and 'big tent' conspiracy theory that pivots around an anonymous figure named Q, who claims that former US president Donald Trump is fighting a cabal of Democratic politicians) created enough emotional pressure to erupt into a shocking real-life event that changed the world. We saw it as well when a group of amateur investors on Reddit took on Wall Street by artificially inflating the price of a failing company, Gamestop. In addition, the memetic warfare driven by 4Chan helped get Donald Trump elected. These are all very different movements, but they all formed online and then breached into the real world.

It's worth noting that breach doesn't have to begin online. Cults have their own shared reality that often clashes with the consensus reality of a society when they try to apply their ideas. Putin's invasion of Ukraine in early 2022 was a breach event. As Andrew Sullivan writes in his essay 'The Strange Rebirth of Imperial Russia,' Putin was heavily influenced by a mythic nationalist fantasy, dreamed up by post-truth Russian intellectuals to cope with the fall of the Soviet Union.[41] That fantasy grew among Russian elites until the pressure built and it breached into the real world, with awful consequences.

Russia's catastrophically botched invasion points to something essential about breach: It never quite goes the way we expected. The virtual and physical worlds are different; they have different rules, different values. When one meets the other, chaos often ensues. Guy Reffitt, one of the ringleaders of the January 6 insurrection, told a reporter from his prison cell that when everything shifted from online and small in-person conversations to insurrection, 'Fantasy slammed into reality like a car wreck.'[42]

We crave community and belonging, so it isn't surprising that many of our new religions and belief systems are forming in online communities. These communities and fandoms can lead us to deep connection and mutual understanding, or dislocation and confusion, or both. And it isn't always easy to find our way through, because certain narratives hook into our emotional needs and our own cognitive biases. When we have a whole community reinforcing those biases, we can quickly fall into groupthink and confusion.

QAnon is one such example; another is the phenomenon of conspirituality, which is the blend between New Age thought and right-wing conspiracy theory that has been developing for decades, and emerged most strongly during the COVID-19 pandemic. As Ward and Voas argue in a 2011 paper in the *Journal of Contemporary Religion*, conspirituality can be defined as:

> ...*a rapidly growing web movement expressing an ideology fueled by political disillusionment and the popularity of alternative worldviews. It has international celebrities, bestsellers, radio and TV stations. It offers a broad politico-spiritual philosophy based on two core convictions, the first traditional to conspiracy theory, the second rooted in the New Age: 1) a secret group covertly controls, or is trying to control, the political and social order, and 2) humanity is undergoing a 'paradigm shift' in consciousness. Proponents believe that the best strategy for dealing with the threat of a totalitarian 'new world order' is to act in accordance with an awakened 'new paradigm' worldview.*[43]

Philosopher Jules Evans has researched this phenomenon extensively, and when we spoke about it, he pointed out that for all its apparent newness, it has deep roots in Christian theology – specifically, the belief that the world is corrupt, a time of reckoning is at hand, and once it comes, the world will be made pure.[44] As Jamie Wheal argues in *Recapture the Rapture: Rethinking God, Sex, and Death in a World That's Lost Its Mind*, this kind of ideology is prevalent in the world right now. While all rapture ideologies are different, what they share is a belief that a time of awakening is coming, and that only a select few who can really see the truth will be allowed to ascend to a new world.[45]

As with all things, there are likely threads of truth woven through all these beliefs. It is important to distinguish between a recognition of corruption in the world and a mistrust of authority, and the more mythical, often paranoid thinking that defines conspirituality. John Vervaeke has suggested that much conspiracy-theory thinking is based on 'the hermeneutics of suspicion.'[46]

Hermeneutics is the study of how we interpret things, and Vervaeke argues that the way conspiracy theory works is by hijacking our sense of revelation – our sense that 'we get it,' that we've tapped into a deep code of how reality works and seen the truth. However, rather than a eureka moment that widens our frames, it is one that narrows our frames because it casts the world as a place of deceit and suspicion that only makes sense through one narrow interpretation. As we'll see in the next chapter, psychedelics can also lead us to these kinds of experiences.

This is not to say that any commentary on systemic corruption should be dismissed as conspiracy thinking. In a world full of institutions and governments that are, quite visibly, broken and corrupt, a healthy skepticism is important. We know from declassified documents that government conspiracies do exist. However, there is a difference between the acknowledgment of actual conspiracies and the dogmatic, almost-religious adherence to a mythical overarching conspiracy. Likewise, it is possible to hold the complexity of our corrupted power systems without collapsing all sense-making down to a single narrative that explains everything.

This is a key theme in many countercultural expressions that percolate online – for example, the alt-right notion that we are living in a 'clown world' posits that we're living in a false reality constructed by the media and a mainstream culture that's trapped in its own fantasy. That same mainstream culture looks back at its critics and declares them insane, as well. On all sides of the political spectrum, we are so trapped in our own reality tunnel that from where we're standing, anyone outside it looks completely insane.

So, perhaps it's no surprise that the sudden awakening from those reality tunnels into the tangible physical realm is incredibly dislocating, whether it's from a psychedelic-induced altered state or a conceptual online rabbit hole we've immersed ourselves in. However, breach has so much psychic force that it often changes the real world when it happens. The protests over the murder of George Floyd took the national conversation around racial inequality in the USA into HR departments, boardrooms, and churches. An issue that people of color had been painfully aware of for too long became a part of the cultural conversation in a new way. On the other end of the political spectrum, QAnon has evolved and morphed from conspiracy theory and proto-religion into a major political force in the USA. As Grid reported in April 2022, 'Grid reviewed public records and reporting, social media posts, and campaign materials and events to identify and confirm at least 78 QAnon-aligned candidates running for office in 26 states in 2022. They're running for governorships, secretaryships of state, seats in the Senate and House, and in state legislatures. They have raised over $20 million this cycle – and over $30 million since 2018.[47]

The rifts between us that lead to phenomena like QAnon are only becoming wider. Until we learn to meet one another in a new way, our social reality is only going to get weirder, more dangerous, and more disconnected.

Flipping, Blending, Grounding, Twisting

Taking seriously the idea that the Internet acts as a kind of collective unconscious means recognizing that our politics and culture are increasingly influenced by non-rational motivations. It means recognizing that in many ways, we live in religious times, and that religious times utilize a different language. It's the language we learn in the psychedelic experience – a language of the transrational, meaning something that includes reason but goes beyond it.

The rise of new Internet religions is a topic I've covered journalistically for a number of years, interviewing psychologists, occultists, philosophers, and religious-studies scholars. In the process, I came up with a model I call psychedelic sense-making, which contains four principles I've found useful in navigating the online realm and psychedelic experiences. They are flipping, grounding, blending, and twisting.

Flipping means becoming adept at jumping between the imaginal realm of the Internet and physical reality, and among radically different possibilities. It involves simultaneously holding cognitive and emotional operating systems at once without collapsing one into the other, something Terence McKenna argued for as a way to simultaneously hold both a Western rational mindset and a shamanic view. Flipping is something we can do in our daily lives to help us think with more flexibility. Flipping has a shapeshifting quality that allows us to explore different ways of being without losing our own sovereignty.

The next principle is grounding. This means staying aware of our bodies and using them as an anchor into the present moment. It is a remedy to the 'luxury Gnosticism' Mary Harrington has pointed to, or the disconnection that is demonstrated in Sherry Turkle's research. It brings us back into the lived reality of our bodies to counter the increasingly disconnected, disembodied virtual world. This could be as simple as getting up and going outside, or taking breaks from the Internet to become aware of our physical sensations. What it shows is that even though it feels like the online world is a space of disembodied minds, there's really no such thing; being online is an embodied process. Embodiment can help us to get better at decentering, or taking a step back from the content of our sensory experience by becoming consciously aware of ourselves right here, right now, in the present moment. The trick is to breathe and learn how to pay attention to our physical sensations. We can also slow down so we don't spin out and attempt to make meaning too quickly, or jump to unhelpful conclusions.

The next principle is blending. Blending means recognizing that seemingly different worlds blend into one another to create something new. It can be as simple as when your work friends meet your school friends, or experiencing intense synchronicities after a psychedelic journey. Instead of fighting or resisting, blending helps us relax into the perfume of the imaginal as it suffuses the world around us. We allow the two realms to blend together rather than burst into one another in opposition. We're living in an era in which the boundary between the real and the imaginal is shifting, just as it does during a psychedelic experience. Reality is not static.

The final principle is twisting. Twisting means changing our own perception to meet what we're perceiving – embracing the weird and the irrational, moving with it rather than away from it. It means accepting reality as it is, not how we'd like it to be, even if it's inexplicable or uncanny. Twisting enables us to hold ambiguity, complexity, contradiction, and weirdness. It is particularly important in finding ways to talk to one another across our reality tunnels.

It lets us meet one another halfway, twisting to accommodate radically different perspectives. It is inspired by a technique developed by the Yaminahua people in the Amazon who drink ayahuasca. As Jeremy Narby explains, the Yaminahua communicate with beings they call yoshi. 'Yoshi are invisible,' he explains, 'and they animate plants and animals. But they are also multifaceted and ambiguous, and all reports of them underline the fundamental difficulty of knowing them...'[48]

If you've ever encountered an online subculture with its own norms, language, and rituals, you might have had the experience of the fundamental difficulty of understanding where its members are coming from. Narby goes on:

> ...the language that they use to sing to these entities is deliberately abstruse and metaphoric; they speak what they call tsai yoshto yoshto, meaning 'language twisting twisting....' Jaguars are called baskets, anacondas are called hammocks, and fish are called peccaries. Anacondas are called hammocks because as they hang from trees they sometimes look like hammocks. Yaminahua shamans say that they use this twisted language to talk to the multifaceted yoshi beings because normal words would crash into them, whereas with twisted words you can go in close...[49]

The Yaminahua deal with radical difference by twisting language to meet it head-on. Similarly, we cannot speak to the transrational nature of our online interactions using the language of reasoned debate we hear on the nightly news. The language crashes into the weirdness of the Internet. Instead, we can find new ways to speak that hold the truth of the space we're communicating in: language that brings us closer together, that allows us to meet those we disagree with as humans with different viewpoints rather than evil people trying to destroy the world.

Revisiting the Intimacy Crisis

Flipping, grounding, blending and twisting are all attitudes that help us lean into the Internet, not unplug from it. Because for all its carnival madness and vice, the Internet may still be the most significant development in the last few thousand years of human history and our best hope for global coordination, meaning-making, and problem-solving.

For the first time in human history, we are connected and can coordinate at scale without the intervention of a centralized power. But this ability is in its adolescent phase, and like adolescents, we have to learn how to be mature and responsible – to relate with virtue, dignity, and compassion.

So far, I have argued that our inability to do this, and the resulting crisis of intimacy, is one of the key drivers of the Big Crisis. It is fueled by our increasing time online, and by the very design of social networks that gamify communication and often prevent meaningful dialogue.

I have also suggested that we need to start making sense of the Internet in a new way to counter this. The Internet need not be viewed as just a technology, but as a distinct realm – much like the DMT realm. It comes with its own rules, its own constraints, and its own emergent intelligence.

As well as recognizing the political, class, and cultural realities that undergird our chaotic online landscape, we also need to take a religious, archetypal, and transrational view. As Hunter S. Thompson said, 'When the going gets weird, the weird turn pro.'[50]

We can't understand our changing cultures through maps and models, or by imagining (as economists do) that society is made up of rational actors seeking out their self-interest. We live in the Age of Breach, in which our unconscious religious impulses take form in the Internet and seek expression in the 'real world.' Psychedelics give us a direct experience of relating to transrational, archetypal realities and navigating them more effectively.

Finding new ways to connect and have authentic, generative conversations about the issues that matter most is more important than ever. Too often, conversations about cultural topics like abortion, climate change, gender, or elections are hijacked and derailed by our outrage, particularly when they're happening online. Thankfully, both modern research and ancient traditions point to the capacity for psychedelic experiences to bring us closer together, and to increase our empathy and capacity to remain open to new perspectives.

Techniques for Connection

Currently, we don't have a lot of research pointing toward how we might use psychedelics to increase connectedness and conflict-resolution processes in a sustainable way. However, as I showed in Chapter One, psychedelics work in combination with a particular practice, ceremony, or ritual. With this in mind, there is tremendous potential to try out different techniques for group bonding with psychedelics. As we'll see in the next chapter, some of these experiments are already happening.

For several years at Rebel Wisdom, we ran hundreds of different online events experimenting with these techniques – without the use of psychedelics. We were hoping to find new practices that could be scaled and applied to overcoming political polarization. Our online gatherings were attended by people from all backgrounds and ages from more than a hundred countries, and there were moments of true magic during which cultural and political barriers were overcome through human connection. In Chapter Two, I explored some first-person practices we can use to navigate the Big Crisis as individuals. Here, I explore some second-person practices that help us relate with one another. And in the next chapter, I explore how they can be combined with psychedelics to move people toward political and social action.

There's an overarching theme that binds all these practices together: It isn't what we say to one another that brings us together, but how we say it. We connect when we actively listen to one another with a positive regard – when we cultivate curiosity and openness. As we have seen, the manner in which we become more receptive during a psychedelic experience may be one reason psychedelics are so therapeutic; it is also a factor in our ability to connect with other people and stay open to one another's differences. When we approach a conversation with these principles in mind, it turns into a beautiful improvisation. New ideas build on one another, and we come up with stunning, unexpected riffs. We have a shared intention, but also freedom within the conversation to move in new directions.

Greg Thomas is an integral theorist and jazz critic who, with his wife Jewel Kinch-Thomas, teaches about how the principles of jazz can help us engage in open dialogue that leads us down improvisational avenues.[51] To engage in mind-blowing jazz improv, every musician needs to have a certain level of skill. In this case, the skills I laid out in the second chapter must be present. These include the ability to regulate our nervous systems and express ourselves clearly and respectfully; the capacity to practice active listening, which facilitator Rich Bartlett describes as 'listening without saying anything, or waiting to say something,'[52] and an authentic curiosity and openness to ideas that may initially make us uncomfortable or unsafe.

One of my favorite aspects of Thomas's model is that it highlights 'antagonistic cooperation,' a phrase coined by Black American scholar Ralph Ellison.[53] This is the idea that we don't have to be in agreement to be creating something beautiful together; the trumpet player might be trying to outdo the saxophonist with his solo, and the bass player might be trying to outdo both of them. But everyone's in on the game and working toward a greater unity in the process.

The mediator Diane Musho Hamilton, in her book *Compassionate Conversations: How to Speak and Listen from the Heart*, suggests that any conflict conversation includes some dance between 'sameness' and 'difference.' Sameness makes us feel safe and connected. Difference can be confronting, but it's also energizing and full of creative potential.[54] Understanding how these aspects work together in our nervous system is enormously helpful if we want to have generative conversations together about complex topics.

Sara Ness, a pioneer in the modality of Authentic Relating, has developed a model called 'context, content, and concern' that can be very useful in any conflict conversation. As Ness and her partner, Geof Krum, explained the model in a course we ran together, 'What we say is the content of our thoughts. However, there is always a shared context going on that shifts over time, changing the frame, goals, and understanding of what the conversation is actually about or for. And there are individual and collective concerns for each of us that define why we're discussing these topics, why we care about them, and why our emotions rise and fall in relationship to each point.'[55] Often, conflict worsens when we aren't explicit about these three aspects by checking in with one another to make sure we are on the same page: Are we sharing the same context? Are we listening to the other person's concerns, and expressing our own? Are we clear on the content of their argument?

To engage in this kind of conversation, we have to understand how our nervous system works. Stephen Porges has argued that our nervous system is actually a social-engagement system. It's constantly responding to other people, telling us whether we can feel safe and receptive or if we need to close up because we're in danger.[56] Staying mindful of our nervous system can help us regulate ourselves, and stay within our 'window of tolerance.' This term is used in the trauma field, and knowing how to work with trauma is another useful tool for having difficult conversations, online or in real life.

A trauma-informed approach to a conversation requires not pushing someone past their comfort zone. All that does is send them into hypoarousal (shutdown) or hyperarousal (fight or flight). One of the best tools I've found for reaching this kind of safety together is through dialogue practices like inquiry, Circling, and Authentic Relating.

These are practices in which we stay aware of the process of speaking while we're doing it. In Circling, for example, people sit around in a circle and speak to what they're experiencing right now. Sometimes they might use a sentence stem, like, 'In this moment, I'm noticing...' and then complete their thought. Or one person in the circle could volunteer to be the center of everyone's attention, asking questions. Someone might say, 'I notice your body language seems tense.' The person might reply, 'Right now, I am feeling tense because you're all looking at me. But saying that out loud, I'm actually noticing my back is relaxing a bit.' Another might comment on this observation. As we speak to what's arising in the present, we unlock new layers of connection and meaning.

When we draw on these and other tools, we can start to combine dialogue practices with making sense of culture and media. One example of this is a process called cultural embodiment, created by Schuyler Brown. During this process, which Brown hosts, a group of people gather on Zoom or in person to watch a controversial piece of media together. However, instead of 'discussing' it, they exclusively pay attention to what they're noticing in their bodies – sensations, emotions, thoughts, and images.

In one session I did with her, she played a clip from a controversial film called *What Is a Woman?* in which the director is having an argument with a doctor around trans issues. After the clip, the group started speaking to what they were noticing in their bodies. People expressed feelings of tension, anger, and frustration. Nobody shared an opinion or a hot take. What was beautiful to notice was that, even though I know for a fact that the group contained people with opposing views, it was impossible to determine anyone's position. Instead, everyone was united in sharing human emotions they were trying to make sense of, and speaking from that place rather than from their opinions.

As Schuyler said at the beginning of the session, she developed the process because, too often in the worlds of spiritual or personal growth, we shy away from the media. We try to shut it out, because often it all feels too base or chaotic. But if we really want to change ourselves, we need to do the opposite by leaning into our discomfort. This is one of the deepest lessons of the psychedelic experience, too – the idea that we need to go 'in and through' whatever is arising to transform and process it.

John Vervaeke refers to the practices I've just outlined as a form of dialogos. This is an ancient Greek process of coming to truth by using language to transcend language. When we gather together to have this kind of transformative dialogue, conversations touch on something deeper than words. Dialogos is like a psychedelic; it unlocks new information, new insights, new ways of seeing together by having a conversation that is more than just a conversation. It takes us past the words to deeper levels of meaning, a space where we're 'communing while we communicate,' as Vervaeke puts it.[57]

What has struck me most in exploring these practices and facilitating retreats is that, when we come together with the intention of having honest, open, compassionate conversations, we can go beyond our own opinions to a much deeper level of shared human connection. And when we really lean in, we can create a collective intelligence between us that is more than the sum of its parts, and that contains within it vast possibilities and creative solutions to our problems.

Closing Comments

We are a social species. Everything we do requires each other. Psychedelics act as powerful tools for social bonding and group cohesion, but we can't rely on the kinds of experiences they elicit to magically bring us together. We can't bypass the gritty reality of life through a mystical experience. Instead, psychedelics invite us to get our hands dirty and deal with the complex reality of political and social division. It's a collective process, but it's also specific to each of us as individuals.

As I lay on the bed while my second dose wore off, I reflected on the spider queen and my other entity encounters. What I was left with was not a metaphysical curiosity, but a burning conviction that it's the bonds between us that really matter. I wanted to change how I showed up in my relationship, and to try to work through my own intimacy issues. That wasn't something the DMT could do for me; shifting would require work, patience, and compassion.

That might be as true on the collective level as it is for each of us in our own relationships with those we love. Relationships take work. Building bridges between different political tribes, or between vax and anti-vax, or left and right, is going to take work.

The political and cultural stuckness many are experiencing is exacerbated by social media, but social media isn't the sole cause. As I have argued in this chapter, the way the Internet is built has a huge effect on how we're able to communicate through it.

If we zoom out, we can say the same about the economic and social foundations of much of the world. The way our systems are built, the incentives they force us to follow may be leading us to ruin. In the next chapter, I explore the darker sides of culture and economics, why they could derail the psychedelic renaissance, and how insights from psychedelic experiences may help us to overcome them.

Chapter Four

Moving Through Darkness

So far, I have looked at how psychedelics can help us widen our perceptual frames on the world, unlocking our mental flexibility and our imaginations. I've explored how they connect us to one another, and to our environment. I've explored the Big Crisis we're facing collectively, and why new frames are precisely what we may need to navigate it successfully.

The thread running through this book has been that psychedelic experiences are transformative. They change who we think we are, what we think we're capable of. But that process is often full of difficulty and challenge. Therapeutically, it often involves wrestling with our demons, twisting and writhing as we unravel profound truths. As psychiatrist Humphry Osmond famously quipped, 'To fathom hell or soar angelic, just take a pinch of psychedelic.'[1] (Osmond is known for coining the term 'psychedelic.') They shine a light on the darkest places of our psyches, and they can help us make sense of the darkest parts of our culture and economy, as well.

In this chapter, I try to unravel what is ignored, hidden, and left out of the psychedelic renaissance. I'll look at the incentive structures that govern our economy, as I offer a critique of psychedelic capitalism and the commodification of the sacred. We'll see how a for-profit version of psychedelic mainstreaming can hinder psychedelics as agents of systemic change, and the ways in which psychedelics can either bring us together or lead us further into our own narrow reality tunnels.

Above all, this chapter is about the tremendous power of the psychedelic experience to reveal what was hidden in our own psyches, and what this can teach us about everything that needs to be made explicit, clear, and conscious in order for us to find new ways to make sense of the world around us.

This has been the most difficult chapter to write. Inevitably, when we delve into the darkness of culture, we encounter our own. There are ideas in the following pages that you may find emotionally challenging. During a psychedelic experience, very often the medicine holds us and gives us the courage to face discomfort; the trust we place in this intelligence, whatever it may be, can see us through. In these pages, things are different. I am far less wise a guide than any of these medicines, so I can only bring my own limited perspective and research to bear in this journey. Luckily, you have yourself – your own sense of discernment, insight, and curiosity – to guide you.

In stories in which someone is compelled to make a journey to a dark and scary place, very often they are given a magical object to help them. So, before we dive in, I invite you to take a moment and bring to mind your own deepest hopes – for the world, for yourself, for those you love. Maybe there is some place in your body where you can feel it – some part of you that feels solid and connected to the enduring power of your human spirit. I invite you to hold it in your palm, carrying it as an ember. At some point you might need it to light the path ahead, because to find a way through the crisis of the times, we may have to descend to the deepest parts of hell.

The Shadow

One of the most significant ways psychedelics can help us make sense of the world is by bringing us face to face with our shadows. The shadow is a concept from Jungian psychology. It is all the parts of our minds that we've pushed away, denied, and locked up in the basement of our awareness. My favorite explanation comes from Zen Master Doshin Roshi, who calls the shadow 'the me I can't see.'[2]

I first learned about the shadow in university while listening to a recorded talk by Ann Shulgin, a pioneering psychedelic therapist. Shulgin passed away at the age of 91 while I was writing this chapter, and while I never met her, I felt her passing keenly, as I'd grown to admire her a great deal. In the talk, Shulgin described how working with the shadow became the most important aspect of her clients' healing in the MDMA sessions she facilitated. Healing took place when they reconnected to parts of themselves that they had previously disconnected from.[3]

Often, our shadows are the 'dark' emotions we've been taught are 'not good,' like anger, greed, and envy. They can also be emotions we'd normally define as 'good,' like confidence or compassion, but that we've disowned within ourselves and projected onto others. Every single person on the planet has a shadow side. But at the same time, none of us like to think of ourselves as the villain, whether we're in an argument with our partner or on trial for war crimes.

Often, our shadows form when we are children, as what Rafia Morgan calls 'an intelligent strategy to survive in the environment you grew up in.'[4] If anger was frowned upon in your home growing up, it could be that any time you had a natural anger response, you stuffed it down so you wouldn't feel it. You had to be seen as a nice person to be safe and accepted. Friendly. Never angry, because anger was seen as wrong. Unevolved. Unnecessary.

So, why can't we just dust our hands at this point, be content in our non-angry ways, and get on with life as the upstanding citizen we are? Even if we acknowledge that we may have parts of ourselves we aren't aware of – hopes, desires, anxieties, dreams, and sexual urges – can't we just ignore them? On the cultural level, can't we just accept that power corrupts, or 'boys will be boys,' and shrug our shoulders? We can and we often do, but there are consequences to our negligence of the shadow, whether it comes in the form of denial or a 'what can I do about it?' indifference. From Ann Shulgin's perspective, 'The shadow aspect of the human psyche, while it remains unconscious, can be blamed for all wars, from tribal conflicts to battles between great nations; it causes racial prejudice; it underlies jealousy and resentment; you'll see the human shadow in every vampire or werewolf movie, and its face is the face of a very popular figure called the Devil, or Satan.'[5]

Carl Jung pointed out that these hidden aspects of ourselves need to be expressed as much as any other part of us, and even when we try to repress them, they still find a way to the surface. We begin to unconsciously project our shadow material onto others and the world around us. If anger is pushed down into the basement, it starts to express itself through how we see others. Maybe other people often seem angry. Or the world itself seems like an angry place. And when we notice that, perhaps we feel contempt. Those angry people might make our lip curl with disgust. Why can't everyone be good like us? Can't they see how *unevolved* anger is? Is it really that hard to keep their anger under control?

And as we're busy issuing these judgments, down in the basement of our psyche, our shadow takes on a life of its own. And any chance it gets, it sneaks out in unconscious behaviors – and until we integrate it, it controls us. It may come out through passive-aggressive notes we leave on the fridge, or road rage. The shadow is that part of us that goes to visit our sick aunt and secretly hopes this will increase the chances she'll put us in her will. It's the part of us that twists the truth to get our way with our partner.

We get into dangerous territory when our projections and our acting out lead us to view someone as less than human; in this case, we're often projecting an unowned part of ourselves onto them. Because it can't be accepted in ourselves, it can't be accepted in the other. As Zen master Doshin Roshi says, in some cases there's 'a hook to hang the projection on.' Maybe they *are* arrogant or bigoted or sneaky. Accepting that we also have those parts in ourselves doesn't mean condoning, or acting on them. It means recognizing that we are human, as Aleksandr Solzhenitsyn wrote, 'The line separating good and evil passes not through states, nor between classes, nor between political parties either – but right through every human heart – and through all human hearts.'[6]

So, if we contain both light and dark, what are we to do with the dark aspects of ourselves? Jung's insight was that they need to be brought back into the wholeness of who we are. We need to accept them, give them a place at the table, and respect the unique perspective they're bringing. As Shulgin explains, 'when work on the Shadow is underway, and it begins to drift towards conscious awareness, it carries with it gifts for its owner…. The Shadow made conscious becomes an ally for us: a fearless, brash, not-quite-housebroken ally and friend.'[7]

Integrating shadow material has been the single most transformative aspect of psychedelics in my own personal healing. There is always more to integrate, but when we bring a part of ourselves back into the whole, we unlock a tremendous amount of psychic energy that we can infuse into our lives. If we integrate our anger, we become more able to channel it in healthy ways – for example, by setting healthy boundaries and speaking up when something isn't fair. When we integrate our vulnerability, we can feel more compassion for the vulnerability of others.

Shadow encounters and integration can be seen as a form of emotional breakthrough, which is a key component of psychedelic therapy, as shown by Leor Roseman et al in a 2019 paper.[8] At the time of this writing, my wife, Ashleigh, was conducting qualitative research at Imperial into how the challenging experiences people have on psychedelics may be a major part of their therapeutic efficacy. Her hypothesis is that the difficulty of coming face to face with difficult emotions that we've been hiding from, processing them, and then recontextualizing them is a key aspect of how we can heal through psychedelics. My own psychedelic experiences, including those I've recounted in this book, have almost always been transformative *because* they were hard, and because they brought me face to face with my shadows.

Cultural Shadows

It isn't just people who have shadows. Cultures have shadows, too: taboos, unspoken histories, abuse, violence, misogyny, racism, greed. Carl Jung believed that 'the collective psyche shows the same pattern of change as the psyche of the individual.'[9]

If psychedelics can help us integrate our personal shadows, could they help us to do it on the collective level? There are a number of group processes aimed at this. One is Thomas Hübl's Global Social Witnessing, which he describes in the following way:

> *Our realization is that collective traumas are at the root of most conflicts, mostly unrecognized and often unconscious. Adequate healing and peace-building is possible only if this is included. First and foremost, this requires the ability to gain a precise and comprehensive picture of what is happening. We call this process of insight, Global Social Witnessing. It is the ability to feel and relate to the cultural process. It's an awareness that the social body is developing through us.*[10]

The proper integration of shadows first involves acknowledging the pain that caused them, and then processing them with great compassion and patience. One example of where this is being trialed with psychedelics comes from a groundbreaking study led by Leor Roseman, a researcher at Imperial's psychedelic research center, my neighbor at the time of writing, and one of my fellow participants on the trial.

Working with Palestinian peace activist Sami Awad and a team of Brazilian facilitators, Roseman observed Israelis and Palestinians who drink ayahuasca together in order to ask what role these ceremonies could play in conflict resolution. The first part of the study, 'Relational Processes in Ayahuasca Groups of Palestinians and Israelis,' was published in 2021.[11] It revealed three key relational themes, suggesting that while the ceremonies helped people connect on a deep level, it also highlighted their differences at times. The ceremonies also included something they called 'conflict related revelations: personal or historical revelations (mostly visionary, but sometimes emotional or cognitive) related to the Israeli–Palestinian conflict.'

The next phase of the study involved taking three mixed groups to Spain for ceremonies focused more explicitly on processing the Israeli–Palestinian conflict. The process involved relational techniques similar to the ones I described in the previous chapter. For example, each participant spent an hour and a half in the center of the group as the main point of focus, answering questions like, 'How does the conflict affect you personally?'; 'When are you passive in your life?'; and 'What are your angels, and what are your demons?' After this, the participant would step out of the circle and the rest of the group would reflect on their answers, inquiring into them as if the person wasn't there. This not only helped to bind the group, but also to surface underlying feelings, ideas, and thoughts around the conflict.

Roseman told me that during this second phase in Spain, everyone had a shared feeling that they were on the cutting edge of conflict resolution and psychedelic healing. The ceremonies were at times chaotic, at times deeply moving, and always complex. Sometimes, people had a shared experience of deep unity. Roseman had just returned from Spain when we spoke, and the study was still ongoing. What follows are his initial reflections prior to analyzing the data.

'One thing many people experienced was this feeling that 'we are the peace,' he explained. 'There was this magical moment in the ceremony – this realization that we are living the peace we are wishing for. You know, it's already happening here for a moment in the circle.'[12]

At other times, the unity of the circle was ruptured as a traumatic memory of the conflict arose for someone, which would create a tension and chaos to be processed and integrated by the group. This complexity speaks to something important about DMT, which is one of the active ingredients in ayahuasca. As I wrote in Chapter One, Rick Strassman has argued that the DMT experience creates a 'prophetic' mystical experience. It's full of insights, ideas, and communication, rather than the type in which our individual egos and desires seemingly dissolve.

Psychedelics and Politics

The mystical nature of the psychedelic experience can have far-reaching ramifications on our 'real-world' interactions. Drawing on Strassman's work, Roseman argues that this has significant implications for ayahuasca's use in conflict resolution. Instead of a bypass, what is needed is, as Thomas Hübl has pointed out, a grappling and processing of collective traumas. Roseman distinguishes between the figures of the mystic and the prophet. The mystic removes themself from the concerns of the world. In some cases, this can lead to bypassing and escapism: an attitude that all is one and conflict is an illusion, so there is no need to engage in it.[13] But the kinds of experiences we can have on ayahuasca or DMT are different. As Roseman explains:

> The prophetic event, to the contrary, is very much political. It is oriented towards others, and it is usually experienced as a conflict – or even crisis – coupled with a tension relief.... The prophetic experience can lead one to become active in fulfilling the ethical–political message that was received in the experience. Therefore, the consequences of it are resistance towards social order.... The 'prophet' sees himself as an active actor on the stage of history, and this requires a creative process – much like that of the artist or the scientist – which engages the entire personality (while the 'mystic' de-subjectifies himself and dissociates himself from personality, space, and time).[14]

Psychedelic ceremonies that elicit prophetic experiences don't remove us from the realities of politics. Instead, they bring them right into the forefront. Roseman told me that the ceremonies were imbued with a kind of revolutionary spirit, a shared desire to actively transform the system. When he shared this, I felt shivers. I told him I felt like he'd just pulled out a stick of dynamite: The combination of psychedelics and revolutionary politics is a powerful, dangerous, transformative proposition. It last erupted in the 1960s and brought with it a burning mass of new ideas, revolutionary energy, suffering, and chaos.

That era of the late 1960s and early 1970s provides a blueprint that can show us why psychedelic ways of being can be both transformative and destructive when they enter a culture that doesn't have a way to contain them. In 1965, the writer Joan Didion traveled to Haight-Ashbury in San Francisco to cover the growing counterculture of young people who were flocking there from all around the USA. In her essay 'Slouching Towards Bethlehem,' she paints a picture of a movement that was at once hopeful and socially conscious, as well as hedonistic and often disconnected from the lived experience of oppression many Americans were experiencing at the time. During this era, a poet named Chester Anderson was seen as one of the people who most had his finger on the pulse of the counterculture. He would print out paper 'communiques' full of updates and general thoughts about the scene, which could be found around Haight-Ashbury. Didion relates one such communique that is particularly revealing:

> *Pretty little 16-year-old middle-class chick comes to the Haight to see what it's all about and gets picked up by a 17-year-old street dealer who spends all day shooting her full of speed again and again, then feeds her 3,000 mikes [of LSD] and raffles off her temporarily unemployed body for the biggest Haight Street gangbang since the night before last. The politics and ethics of ecstasy. Rape is as common as bullshit on Haight Street. Kids are starving on the street. Minds and bodies are being maimed as we watch, a scale model of Vietnam.*[15]

The horrific sexual and drug abuse that were prevalent at the time are often overshadowed by the popular conception of the psychedelic 1960s as a utopian era defined by peace and love. We have to be careful not to confuse a revolution in consciousness with a revolution in society, although the themes of both often converge.

How they converge is often complex. Ram Dass, Alan Watts, and other figures of the time opened people up to a whole new way of conceiving of themselves by bringing Eastern spirituality into the West. At the same time, Timothy Leary was known as 'the most dangerous man in America,' in part because he was telling young people to 'drop out' of their culture through a radical shift in consciousness.

While these new spiritual ideas certainly had an impact, it's hard to separate them from the political realities and social movements of the 1960s. During this era, 'the most dangerous woman in America' was Bernardine Dohrn, the leader of the revolutionary group known as the Weather Underground, colloquially called the Weathermen. They were committed to violent resistance and sabotage to end the war in Vietnam, and allied with groups like the Black Panthers to fight white supremacy and racism in the USA.

They didn't have particularly strong links to the psychedelic counterculture, but they did break Timothy Leary out of prison in 1970, helping him to find temporary asylum in Algeria with the Black Panthers. As reported in the podcast 'Mother Country Radicals,' Black Panther leader Eldridge Cleaver grew critical of Leary and his obsession with a revolution in consciousness, instead of a political revolution to overthrow a system defined by imperialism and racial oppression. In a statement to the Weathermen, he expressed:

> *We took upon ourselves Timothy and Rosemary [Leary's wife]*
> *at your request, in order to demonstrate our love and solidarity*
> *for you and our great, undying respect for your beautiful*
> *revolutionary work. But we also want to say that it's become clear*
> *to me there is something seriously wrong with both Dr Leary and*
> *his wife's brain. And I attribute this to the multiple acid trips*
> *they have taken.*[16]

It is not hard to imagine why a man like Cleaver, who was at the forefront of a battle against racial inequality that was destroying his community and that had seen friends and comrades killed on a regular basis, would feel frustrated by the insistence of a rich white psychiatrist that an abstract revolution in consciousness would fix everything. This points to something Roseman noted in the first phase of his study: There was a tension between spiritual conceptions of oneness and the lived reality of separation and warfare; and both had to meet in a real, human, and complex way for any significant change to occur. Leary would eventually be arrested again, and go on to rat out the very people who had broken him out of jail in order to reduce his own jail sentence.

A revolution in consciousness is only skin-deep unless it's tied to virtue and integrity, and psychedelics have the potential to enhance narcissism and grandiosity as much as they do to reduce it. Despite these complexities, we shouldn't throw the revolutionary baby out with the psychedelic bathwater. Instead, we might learn from the past to mitigate these kinds of horrors. As we'll see in the next chapter, psychedelics do have the potential to play a revolutionary role in society if approached with respect and reverence. And it may be that the most profound revolutionary act is not undertaken by flipping the world upside down, but by taking care of one another.

One of the most memorable parts of my conversation with Roseman was listening to him describe how some of the participants had decided to move to the same towns when they came back. They looked out for one another, even helping each other move house. Despite this deep human connection, many also struggled to reconcile the deep understanding and connection they'd felt together with the cold reality of the political situation. This speaks to something crucial about the use of psychedelics. No matter how much we expand in the moment, we always have to return to the realities of the systems we live in.

The God of Destruction

Studies like Roseman's open up an interesting question: Can our individual processes of transformation, or work in small groups, ultimately change entire systems? This idea – that if enough individuals grow up and change, the system will automatically change, too – is popular in New Age and systems-change circles. It is probably partially true, but incomplete. As we saw in Chapter Two when we examined the Big Crisis, we are all part of many overlapping, complex, adaptive systems. The system responds and defends itself as if it were a living thing. And just as our social-media networks are designed to force particular values onto us, so too do our cultural and economic institutions.

To have any hope of collective transformation, psychedelic or otherwise, we must face what it is about our social and economic systems that resist change. To explain why this is so important,

I'm going to summon a god. Normally this would be a risky thing to do, because gods, once summoned, can't be controlled. But in this case, the god is already here with us.

The god is called Moloch, and he is a god of war to whom the ancient Canaanites used to sacrifice their children. In exchange for what is most precious to us, Moloch grants us power in the world. Moloch was revived by Allen Ginsberg in his poem 'Howl,' which casts the god as an ever-present and unseen force driving the world toward disconnection:

> *Moloch the incomprehensible prison! Moloch the crossbone soulless jailhouse and Congress of sorrows! Moloch whose buildings are judgment! Moloch the vast stone of war! Moloch the stunned governments!*
>
> *Moloch whose mind is pure machinery! Moloch whose blood is running money!*[17]

Ginsberg links Moloch to capitalism and the extractive, heartless pursuit of progress. But it's also more than that; it's a kind of force that people have given many names to throughout history. It's a metaphor for our own capacity for blind extraction, and what happens when we create a system around this blindness that takes on a life of its own. Moloch is a powerful metaphor because it casts our economic system as a single entity with a motivation of its own. It's an oversimplification, but it makes it much easier to make sense of the world.

To see why, we can look to essayist Scott Alexander; in his groundbreaking essay, 'Meditations on Moloch,' he re-popularized the concept by explaining it through game theory: the branch of mathematics that analyses competition.[18] For

Alexander, Moloch is the perfect metaphor for the incentive structures within our economy that lead to race-to-the-bottom situations that leave everyone worse off.

Former poker pro and YouTuber Liv Boeree summed it up clearly when we spoke: 'If there's a force that's driving us toward greater complexity, there seems to be an opposing force, a force of destruction that uses competition for ill. The way I see it, Moloch is the god of unhealthy competition, of negative sum games.'[19]

In Chapter Two, I described how complex adaptive systems lead to emergence—that incredible, magical quality that sees novelty growing organically from the interaction of many parts. Moloch can be seen as the opposite of emergence. Moloch is entropy. Decay. Emptiness. Disconnection. As Scott Alexander argues, an essential concept for understanding how this plays out in society is the 'multipolar trap.' This is the idea that when individuals are given the incentive to do something that is detrimental to the group as a whole, everything eventually goes to shit for everyone.

Imagine a pristine lake. There are 10 fish farms on the shore, separated from the rest of the lake but using the same water. To prevent pollution, each farm has a sophisticated filter. For a while, all the farmers use the filter and the lake stays pristine. Then, one day, one of the farmers finds a way to run her farm without the filter, without any regulators noticing. She spends many sleepless nights ruminating over the decision, but the farm hasn't been profitable and she has bills to pay. She can only survive by paying bills, which means she has to keep farming fish. She can only keep doing that if she turns the filter off. She does so, reluctantly.

Without the additional cost, her profits start to soar. In fact, she does so well that she's able to buy the farm next to her, which has also been struggling. Soon, she owns three farms. The rest of the fish farmers soon realize what she's doing, but if they go through the bureaucracy of reporting her to regulators, they might end up going out of business. Plus, she now has enough money to hire expensive lawyers. And so, they have a choice: cheat or perish. They all have families to feed, so they cheat. Soon, nobody is using a filter and the lake is completely polluted. Shortly after this, the whole ecosystem collapses. The lake can no longer support fish farming. The remaining farmers move on to another lake. Moloch cracks his knuckles.

This is one example of a multi-polar trap. We can see it everywhere in a consumer culture that values acquisition and progress above everything else. It is important to frame this force as a god, because we have come to worship it. As historian Eugene McCarraher argues in *The Enchantments of Mammon*, capitalism has come to replace religion.[20] As summarized by Daniel Steinmetz-Jenkins in The Nation, McCarraher argues that:

> *...the mysteries and sacraments of religion were transferred to the way we perceive market forces and economic development... a 'migration of the holy' to the realm of production and consumption, profit and price, trade and economic tribulation. Capitalism, in other words, is the new religion, a system full of enchanted superstitions and unfounded beliefs and beholden to its own clerisy of economists and managers, its own*

iconography of advertising and public relations, and its own political theology.[21]

Moloch isn't just a force. It is the religious substructure of any country that sees economic growth as the highest value, which is most of the world right now. And as Scott Alexander puts it, 'Only another god can kill Moloch.' But we don't have another god in capitalist societies. And this presents a tremendous problem for the mainstreaming of psychedelics.

Psychedelic Capitalism

As we've seen, psychedelics elicit sacred experiences that connect us to values beyond profit, acquisition, and perpetual economic growth. The psychedelic counterculture for the last 50 years has been, to varying degrees, committed to a view of the world that is more oriented around what is beyond the physical world.

This has led to an idea among researchers and activists that I call the 'psychedelic Trojan horse.' This is the notion that psychedelics can be snuck into our existing system in the container of medicalization and for-profit pharmaceuticals. Once that happens, the thinking goes, the transformative power of the molecules will transform society. I have heard some variation of this idea from many of the most prominent scientists and activists who ushered psychedelics through the 40 years of prohibition to bring them to the cusp of legalization.

In some ways, the argument holds; the psychedelic experience has a wisdom and teaching of its own, one vastly wiser and more powerful than we really understand. But it always teaches in synergy with our own perceptual framework and the cultural stories we have around the experience. As we encountered in the first chapter, psychedelic experiences are heavily influenced by our set, setting, and dose. This is true of our cultural setting as much as the room or field we happen to be in; when you have a psychedelic experience, you're bringing your whole culture with you.

As Erik Davis explained to me, the cultural narrative we have around psychedelics *changes the experience* we have of them.[22] Because of this and other factors, the psychedelic Trojan horse argument is a fallacy. When it comes to psychedelics, the horse is as important as what's inside it. It's a folly for psychedelics to enter into our culture's existing narrative through a for-profit medicalized model that doesn't recognize anything beyond itself – a model centered around Moloch's endless quest for profit. Instead, they need to enter *with a new narrative* – one that moves beyond profit. In a capitalist system, even human life is secondary to the profit motive. As economist Atul Gupta revealed in a disturbing and extensive 2021 study, when private equity firms take over nursing homes in the USA, deaths increase by an astonishing 10 percent, due largely to the staff cuts they make to boost profits.[23]

But the cat, or the giant horse, is already out of the bag. At the time of this writing, there are just under 50 psychedelic pharma companies on stock markets around the world. Each of them is incentivized to compete in the same way the aforementioned fish farmers are, or the private-equity firms who own nursing homes. The medicalization landscape is complex, and there are also nonprofit organizations like the Multidisciplinary Association for Psychedelic Studies (MAPS) and Usona Institute, who are successfully exploring alternative models.

Also at the time of writing, MAPS is on the cusp of legalizing MDMA as a medicine, and Usona is in Phase III trials for psilocybin as a treatment for depression. However, as MAPS founder Rick Doblin shared when we spoke, the presence of for-profit companies is making it harder for nonprofits to raise money. Philanthropists can now make money instead of giving it away.[24] Moloch smiles.

Other alternatives are decriminalization movements that are underway – for example, the US movement known as Decriminalize Nature, which was instrumental in sanctioning psychedelics in the city of Oakland, California. On a larger scale, the state of Oregon successfully legalized the therapeutic use of psilocybin in 2021, with a full rollout expected in 2023. In the Oregon model, treatment and access will be regulated by the state, and at the time of writing, it seems they are likely to adopt a two-tier model, with clinical treatment for those with mental-health conditions, and at retreat centers or one-to-one sessions for non-clinical populations. Psilocybin and other psychedelics can be grown extremely cheaply, so there is no need in this system or other models for expensive pharmaceutical versions of them.

The question remains as to whether the structures of market capitalism will be incentivized to attempt shutting efforts like this down. There is a lot of money at stake, and psychedelics are becoming big business, with the psychedelic pharma industry predicted to be worth $8.31 billion by 2028.[25] With investment comes opportunity, but also the multi-polar traps of capitalist market economics.

A good example of this is the controversy over patenting that hit headlines in early 2020 and is still raging at the time of writing. As psychedelic pharma companies started up, they began filing what many saw as overbroad patents with the intent of cornering the market. While this is a regular practice in the pharma and tech industries, I and many others saw this as a worrying sign of how psychedelics are entering culture.

To understand why, we can take a brief look at the reason these companies need patents. It isn't possible to patent a psychedelic molecule like psilocybin, as it already exists in nature. This causes an issue for pharma companies, as it's hard for companies to raise any money without being able to show that they will have some kind of exclusivity over what they produce. What can be patented is a particular synthesis method, and arguably, a protocol for using a psychedelic to treat a particular disorder, such as eating disorders or anxiety.

In 2020, the leading psychedelic pharma company, COMPASS Pathways, caused controversy when a patent filing emerged that included what many saw as overly broad patents aimed at shutting down the competition. In the filing, the company appeared to be attempting to patent minute aspects of the psychedelic experience, including a therapist placing their hand on the patient's shoulder, and 'soft lighting' in the dosing room.[26] I covered this journalistically at the time, and in May 2020, I had a public debate with Lars Wilde, co-founder of COMPASS Pathways.[27] In it, I argued that the aggressive patent strategy COMPASS and other psychedelic pharma companies are taking is not only unethical, but will kill innovation in the field by shutting down the competition.

I also made the point that what is needed is a healthy psychedelic ecosystem, which includes clinics as well as churches, many different ways people can access psychedelic experiences, and a series of checks and balances – much like the US constitution, which ensures that no single 'species' (like a pharma company) can become invasive and take over the whole forest. Added to this, there are other ways for companies to be rewarded for the risks they take. Rick Doblin told me that data exclusivity is one viable option. It allows companies to own their data for 10 years, but also enables others to use the molecule or process in question.[28]

Wilde's counter-argument was that COMPASS was pursuing a strategy of patenting because it's the only way to ensure that essential medicines get to people who desperately need them. Without the patents that gave them exclusivity, they couldn't raise money and thereby offer the medicines to more people. He maintained that COMPASS wouldn't try to shut down the competition or prevent people from growing mushrooms for personal use, but my argument was and remains that Moloch gives them no choice but to.

There is no reason why, once psychedelics are legal medicines, pharma companies will be incentivized to cure illness; it's more likely that they will find a way to keep people coming back for dosings as frequently as possible, or continuously upsell new psychedelic treatments. Some companies are trying to mitigate this kind of scenario with alternative business models. One venture capitalist fund, Vine Ventures, reinvests

50 percent of its profits in philanthropic causes within the psychedelic sphere, while companies like Woven Science invest in indigenous reciprocity initiatives.

Vine Ventures co-founder Ozan Polat told me that the key difference in their philosophy and structure is a focus on distributing wealth rather than maximizing and then hoarding it.[29] In a similar vein, MAPS has a for-profit 'public benefit corporation' that is wholly owned by the nonprofit. Instead of a mission to maximize profit, it is set up to benefit the public good. Speaking at the Interdisciplinary Conference for Psychedelic Research (ICPR) in 2022, Doblin received applause for going on to suggest that all pharma companies should be using this model.

These are important efforts, but they are pushing against powerful incentive structures. As Jamie Wheal pointed out in a series of films I made around this topic, many researchers and academics in the legacy psychedelic community seemed naive to the game-theory dynamics of capitalism when they first started engaging with venture capitalists and pharma execs. Many for-profit players came into the ring telling touching stories about their own psychedelic experiences and how much they cared about healing the world. As Wheal explained: '[The narrative of] I've looked into their eyes, I've seen their soul, we've broken bread, we've tripped for a night and they've told me their "profound journey" story which brought them to this "space"? All those things go out the fucking window when we have the multi-polar trap.'[30]

What is needed, as many are calling for in the psychedelic space, is a healthy ecosystem that isn't dominated by for-profit pharma companies and venture-capitalist firms. Having researched psychedelic capitalism extensively, I believe the only way to do this is through a change at the level of the system itself. It's important to note that this doesn't mean there aren't people within for-profit companies who care deeply about social change – there are. What it means is that it's hard for their values to outcompete the system. Just as oil companies 'greenwash' themselves to appear environmentally conscious, or make their logos rainbow-colored during Pride Month, psychedelic companies are incentivized to 'tie-dye' their image as caring and conscious, for PR benefits. Whether it comes from an authentic desire or not is largely irrelevant if they're forced to play a race-to-the-bottom game.

Somehow, psychedelics need to enter society in a cultural container that allows for a paradigm shift at the level of what we value. Paradoxically, a capitalist version of psychedelic mainstreaming could perpetuate the harmful economic, social, and mental-health paradigms that underlie the mental-health crisis.

Access and Reciprocity

There are concerted efforts by many psychedelic researchers, clinicians, pharma companies, and activists to try and ensure that, as psychedelics go mainstream, they do so equitably. One key question is how to ensure that the psychedelic renaissance welcomes as many people as possible from different socioeconomic backgrounds. It costs tens of millions to bring a pharmaceutical drug to market, and companies offering psychedelic therapy are under pressure to recoup that cost. At the time of this writing, a course of ketamine therapy in the UK costs around £6,000, placing it well beyond the affordability of most people.[31]

If people are forced to access psychedelic experiences through a purely medical model, many people won't be able to afford to; others won't feel welcome. As mentioned previously, the city of Oakland, California, became the first city in the USA to decriminalize psychedelics, in 2019. Nicolle Greenheart is the co-founder of the Decriminalize Nature movement that was instrumental in the campaign and works as a facilitator within the community, which has traditionally been home to many African Americans.

As she explained to clinical psychologist Monnica T. Williams: 'I could completely relate to the belief that psychedelics were just White people hippie drugs tied to the wild and crazy 60s.... I know what it's like to be the only Black person in the room talking about the healing power of plant medicine and wondering how I can get more of us not only in the room but truly benefiting from the medicine in a way that feels safe and honors our own ancestral roots.' As Williams goes on to argue, many people of color whom she has spoken to have good reason to feel mistrustful of medicalized psychedelics, and are reluctant about the vulnerability that accompanies being in an altered state. Despite this, she argues that 'these medicines are part of [a Black American] cultural birthright, and I believe we lose more when we step back and choose not to engage.'[32]

What Williams and Greenheart are both pointing to is that the space of psychedelics thus far has been very white and middle-class. Models of care and access are created through a lens that doesn't always recognize different needs or approaches to healing other communities. For an equitable, diverse, and vibrant psychedelic future, that has to change.

Truth and Lies

As well as issues around access, another shadow of the psychedelic renaissance is the downplaying of the dangers. Rosalind Stone is a writer and media professional who has headed up press for the Beckley Foundation and the Conservative Drug Policy Reform Group. She told me that 'feature pieces on psychedelics are now so prevalent as to feel ubiquitous, but they tend to extol their potential benefits at the expense of critical engagement with the data. It's an inversion of the apocryphal headlines claiming LSD makes those who take it infertile in decades past, but no more nuanced and less hyped.'[33] And while newspapers are happy to talk about scientific benefits, it's extremely rare to see articles arguing for a change in drug laws, or anything that might seriously challenge conceptions around how society should function.

Philosopher and writer Jules Evans created an organization called Coming Home to conduct research into the challenging experiences people have on psychedelics to contribute to a more realistic and nuanced understanding of them. As Jules told me, the organization was set up in part because these risks are so under-researched, to begin with.

As well as psychological dangers, psychedelic experiences may also present physical risks. Journalist Ed Prideaux works with the Perception Restoration Foundation, a charity that raises awareness of hallucinogen persisting perception disorder (HPPD). The charity's website describes it as a neuropsychological condition that causes sufferers to perceive a world obscured by 'thousands of blinking dots, shimmering flashes of light, and other perceptual changes that overtake their life.'[34] Prideaux told me that it often results from traumatic psychedelic experiences and is usually accompanied by anxiety and depression, as sufferers feel alone and helpless. Prideaux believes it's both under-reported and under-researched in the psychedelic renaissance.

Anna Lutkajtis, author of *The Dark Side of Dharma: Meditation, Madness and Other Maladies on the Contemplative Path*, has argued that we've seen something similar happen in the way the West has adopted meditation practices. Often, they are viewed as 'all good,' while the downsides of meditation, such as derealization or difficulties processing intense spiritual or emotional breakthroughs, are ignored or brushed over.[35] When we spoke, she shared that when the book came out, she received dozens of abusive emails from people within the spiritual and meditation worlds. Many were outraged that she was suggesting their practices could have any serious downsides.[36] This points us back to the shadows that can fester around our spiritual traditions when we fail to recognize that they aren't perfect.

Delusion

Speaking to Prideaux and Lutkajis, I recalled my own experiences of HPPD after a difficult trip in my mid-twenties. I experienced years of anxiety, panic attacks, facial apophenia (the tendency to see facial patterns that are not truly present), and brain fog. Much more significant was the intense existential crisis I experienced that is sometimes called 'spiritual emergency.' It would ultimately change my life in significant and positive ways. However, it's not an experience I would wish on anyone, and I was lucky to have enough support and knowledge to help me get through it. I have known people who attempted suicide and sometimes succeeded in taking their own lives after bad trips.

This speaks to something that is essential to understand about psychedelics. They are transformative partly *because* they are dangerous. I see them as the psycho-spiritual equivalent of skydiving. They give you an incredible level of insight into the world, spreading out a vast landscape of awareness and clarity. But if you aren't careful, prepared, and humble, there's a chance you won't make it back.

Or you might pick up some unhelpful ideas on the drop. In the last chapter, I examined John Vervaeke's concept that we can have revelatory experiences that narrow our perception instead of widening it. Psychedelics have an effect on the salience network in our brains, and the experience is often one in which everything feels more meaningful than in normal consciousness. We see patterns everywhere and make connections where there may be none.

Without keen discernment, and a supportive community who can give us a reflection on our new perspectives, we can easily fall into rabbit holes. We can fixate on ideas, ideologies, fantasies, or paranoid stories in such a way that we narrow our frame rather than widen it. It's also unclear as yet whether psychedelics lead us to a more compassionate and interconnected worldview; it could also be that they are 'values-neutral.' Researchers Brian Pace and Neşe Devenot argue in a 2021 paper that psychedelic experiences are used by the far right to promote hierarchy-based systems, suggesting that 'any experience which challenges a person's fundamental worldview – including a psychedelic experience – can precipitate shifts in any direction of political belief.' They go on to suggest that 'historical record supports the concept of psychedelics as "politically pluripotent," non-specific amplifiers of the political set and setting.'[37]

As well as being complex in how they can change our minds politically, they also require a lot of care and support to heal. As psychedelics become more popular, more people are seeking out underground therapy because legal therapy is so hard to access. While there are some excellent underground therapists with a lot of integrity, there are also a lot of unqualified or under-prepared people attempting to hold space for others in a therapeutic or ceremonial context.

I spoke about the consequences of this with Pamela Kryskow, the medical lead at a nonprofit Canadian psychedelic clinic called Roots to Thrive. The clinic has been facilitating group psychedelic-assisted therapy with ketamine and psilocybin, mainly to veterans and first responders, to treat PTSD, depression, OCD, and other illnesses. She told me that their team has had to rescue and help people who have received underground forms of therapy. 'Some landed in the hospital due to poor care in dosing of psychedelic medicines,' she explained, while others 'have felt destabilized by unskilled, or lack of, integration following the medicine session. There are many skilled therapists working in the underground; however, without oversight of professional peers, the unskilled ones cannot be stopped.' She pointed out that it's also rare that an underground therapist has the medical understanding to know how psychedelics will interact with different medications.[38]

Highlighting these risks and complexities is not to say that psychedelics aren't safe and effective overall when taken in a conducive setting with adequate preparation, integration, and access to spiritual or psychological support. Nor is it an argument for a fully medicalized psychedelic future. However, brushing their dangers under the carpet in the fear that talking about them will derail the psychedelic renaissance is misguided. A psychedelic future founded on unresolved shadows is, in my view, not worth having. If we don't acknowledge the dark side of these powerful medicines and honor them within the narrative, we are doing nobody a service.

Indigenous Reciprocity

Another key concern for many involved in psychedelics is how to give back to the people whose medicines, practices, and worldviews are being used and exploited for profit. As Dennis McKenna told me during a conversation around psychedelic capitalism:

> I'm not anti-capitalist, but I am pro-ethics. Here's the issue: If psychedelics are going to be integrated into mainstream society, we have to recognize their origins are in indigenous societies. And we owe a big debt to them for being the stewards of this knowledge for essentially thousands of years…. We need to acknowledge that debt.[39]

There are several organizations focusing on what this could look like. On a panel we were on together at Breaking Convention,[40] anthropologist Gabriel Amezcua said something I think is essential to hold in mind in this conversation. Reciprocity is a process of mutual respect and exchange. So, instead of assuming all psychedelic-using indigenous people are one homogeneous mass, we should recognize their diversity and actually ask different groups of people what they want.

Reciprocity can quickly become another form of colonialism if it is framed primarily around former colonial powers deciding how the reciprocity should look, and enacting it in a way that doesn't challenge the same systems that benefited from its absence. Once again, this makes it all about us. Instead, we can move toward a healthy, respectful, reciprocal relationship between cultures that still recognizes the power dynamics at play. That process begins with listening.

The Chacruna Institute, run by anthropologist Bia Labate, has been wrestling with this question for years. Their approach to indigenous reciprocity sees them acting as a connector and supporter of various indigenous organizations they spent several years finding and building relationships with, in order to understand their unique needs and perspectives.

This can also happen in the for-profit space. The psychedelic company Woven Science dedicates 10 percent of its profits to a reciprocity project called El Puente. One initiative this has supported includes partnering with indigenous businesses like Kené Rao, an intellectual-property defense fund and mini-factory run by Shipibo-Conibo people in the Peruvian Amazon.

Kené Rao is led by Demer Vasquez, who is one of only three Shipibo-Conibo lawyers in the world. When we spoke, he told me that, from his perspective, medicines like ayahuasca, which are sacred to the Shipibo-Conibo, can't be separated from the cosmological framework around them. Even though my Spanish is rusty and I had a translator on the call, Vasquez struck me as an activist lawyer through and through: passionate, eloquent, and convincing. He explained that Kené Rao wants to protect the traditional art, the Kené, as the intellectual property of the Shipibo-Conibo, along with having a say in how ayahuasca enters the market. This may include a right to some of the profits that come from this cultural IP, which would be funneled back into the community. In some ways, it's the Shipibo-Conibo equivalent of the Champagne region of France owning the rights to their unique brand of wine.[41]

Kené Rao are trying to play the game of market capitalism to protect their culture and communities. As a charity, Chacruna is trying to establish a fair and methodical form of reciprocity to help grassroots communities. As important as their work is, both still rely on the wider incentive structures of our current economic system. Chacruna needs donations from organizations within the economic system, while Kené Rao needs to compete within it. Neither are truly free from Moloch.

This matters a great deal, because the most diabolical aspect of capitalism is that it can take anything that might pose a threat and turn it into a weapon. While many psychedelic pharma companies talk about their commitment to access and reciprocity, we have seen that before. I would argue that as long as psychedelics are entering Moloch's world, they will share the same fate as other practices, ideas, and perspectives that can change the system.

When Social Justice Gets Appropriated by Moloch

One striking example that illustrates how transformation can be captured by Moloch is the way in which social-justice movements have been adopted by corporations, and the resulting culture wars that such appropriation has fueled.

It should be noted that, once again, we are venturing into complexity. Any maps I am drawing are drawn on shifting sands, and there will inevitably be important perspectives I miss. However, exploring how our existing system hijacks anything that could threaten it is crucial to understanding whether psychedelics can, in fact, lead to significant social change.

Many of the culture wars in the West center around the concept of being 'woke,' a term that initially referred to the importance of staying awake to the very real, sometimes obvious, but often subtle nature of systemic oppression in society. In particular, it arose from critical theorists who were pointing to the ways in which systems of power disproportionately disadvantage minorities and anyone whose identity doesn't fit into the dominant narrative of their culture. This perspective has its roots in postmodern theories around how power works, and deconstructionism – the idea that truth is socially constructed, and because of this, our social norms aren't set in stone; instead, they are made up by the stories we tell.

These critical theories are pointing to something very real. According to a UK government website, 'Black people are over four times more likely to be stopped and searched by police as the national average, and seven times more likely than white people.'[42] Only 4 percent of FTSE 350 CEOs are women.[43] The richest 10 percent of the global population currently have 52 percent of the income.[44] There are very real, deeply rooted inequalities in developed societies.

Despite the important critiques it brings, something changed when social-justice theory was adopted by corporations and governments. On the one hand, it has brought overlooked issues from the fringe into the mainstream, but on the other, it has become a tool wielded by the very power structures it seeks to deconstruct.

Civil-rights activist and radical-feminist author Audre Lorde highlights this dynamic best in her essay 'The Master's Tools Will Never Dismantle the Master's House,' in which she writes, 'What does it mean when the tools of a racist patriarchy are used to examine the fruits of that same patriarchy? It means that only the most narrow parameters of change are possible and allowable.'[45]

One result of this is that, as social-justice ideologies are entering workplaces, media, and government, they are presented as a tightly controlled orthodoxy that can't be challenged, instead of an idea that can transform the system. Change is tightly controlled, and any deviance from orthodoxy is punished at the level of social status.

In much the same way, we might end up with a corporate-controlled psychedelic renaissance, where pharmaceutically based medical access is the only kind available. If this happens, it is likely that any conversation around social change will be tightly controlled to ensure it doesn't affect anyone's bottom line. We can see a version of this already, as corporations are desperate to appear to be bending over backwards to show just how diverse and caring they are.

But in many cases, PR releases and advertisements don't line up with action. When social critique threatens profits, it all goes out the window. For example, many companies have come under fire for ditching their values to appeal to the Chinese market. When Disney launched *Star Wars: The Force Awakens* in China, they significantly shrank the black actor John Boyega in the poster and faced accusations of racism. As Boyega expressed in a GQ interview in 2020: 'What I would say to Disney is do not bring out a black character, market them to be much more important in the franchise than they are, and then have them pushed to the side. It's not good. I'll say it straight up.'[46]

Similarly, Starbucks prides itself on being an inclusive workplace, and has attracted many LGBTQ+ employees. At the time of this writing, many Starbucks employees in the USA are unionizing and facing intense resistance. As reported by *The New Yorker* in the summer of 2022, 'Starbucks has fired some two dozen organizers and stalled union votes by contesting the legality of individual elections. It has promised more pay and benefits to employees at non-union stores and shut down several unionized locations.'[47] Speaking to *The Times* in 2022, CEO Howard Schultz said that he would never accept a union at Starbucks.[48]

Corporations are often happy to champion diversity and inclusion, but not if that involves a redistribution of power and wealth. Marxist scholar Adolph Reed has argued that identity politics is simply another tool of capitalism.[49] Institutions focus so heavily on identity issues because it allows them to redirect the conversation from the underlying causes of inequality, which are primarily about socioeconomic class. Solving class issues requires genuine systemic change. Endless arguments about identity do not, and are therefore selected by elite institutions in order to maintain the status quo.

All of this reveals a deeper shadow in our economic system. The largest threat to Moloch would be a serious challenge to the corruption of the financial and political systems that lead to racial, gender, and economic inequalities. This would have to entail a concerted effort to address growing wealth inequality, lack of class mobility, environmental degradation and widespread corruption – not at the surface, but from the very roots of the problem: at the level of the ones who have economic power and agency. For beneath this is the deeper level of how we perceive, how we think, and what we believe we deserve.

Psychedelic Class Wars

This brings us back to the question of access. If psychedelic therapy can only be accessed by the educated and wealthy, it could deepen the class dynamics and inequalities that exist in most developed nations. In a worst-case scenario, the rich gain powerful transformative personal experiences that don't extend beyond self-interest, while the poor are arrested for growing their own mushrooms.

A renaissance for the rich would heighten not just the economic divide, but the cultural divide, too. In *The Road to Somewhere: The Populist Report and the Future of Politics*, British journalist David Goodhart has highlighted a growing divide between 'somewheres' and 'anywheres.' Anywheres are cosmopolitan city dwellers who participate in a globalized knowledge economy, usually doing jobs that involve producing ideas or working with information. They tend to be progressive and don't feel deeply connected to a particular place. They are as comfortable in a New York coworking space as they are in a hotel in Paris. Somewheres are more locally rooted – in their state or region, for example – and tend to skew more conservative.[50] Many commentators have attributed the Brexit vote in the UK and the election of Donald Trump to this increasing split between a metropolitan elite, who sneer at the Somewheres, as Hillary Clinton did with her infamous line that Trump's supporters were 'a basket of deplorables.'

More recently, the writer N.S. Lyons has suggested a similar split between what he calls 'virtuals' and 'physicals.' Virtuals typically work white-collar jobs that can increasingly be done online. Physicals do blue-collar jobs that involve their bodies, jobs that enable the virtuals to live their lives: They drive trucks, work farms, and keep society running. Lyons argues that our culture wars are intensifying because the two groups have become increasingly disconnected from one another.[51] He is partly drawing on the work of historian Christopher Lasch, who predicted these trends in his 1997 book, *The Revolt of the Elites and the Betrayal of Democracy*, writing:

> *The thinking classes are fatally removed from the physical side of life.... Their only relation to productive labor is that of consumers. They have no experience of making anything substantial or enduring. They live in a world of abstractions and images, a simulated world that consists of computerized models of reality – 'hyperreality,' as it's been called – as distinguished from the palatable, immediate, physical reality inhabited by*

ordinary men and women. Their belief in 'social construction of reality' – the central dogma of postmodernist thought – reflects the experience of living in an artificial environment from which everything that resists human control (unavoidably, everything familiar and reassuring as well) has been rigorously excluded.[52]

Could psychedelics reconnect physicals and virtuals? With complex roots stretching into shamanism, psychiatry, rave culture, ceremonies, religions, and the revolutionary underground, psychedelics straddle many different classes and backgrounds. The potential exists, as long as psychedelic mainstreaming involves more than just for-profit medicalization. If it does, there is tremendous potential for processes and protocols that use psychedelics to start healing these deep rifts.

The Myth of a False World

We can see how psychedelics could heal some of our cultural rifts by looking at a curious kind of myth that's forming on different ends of the political spectrum. It's a myth that has been intricately interwoven with psychedelics throughout history, and it's one that appears over and over again. We can recognize some threads of it by looking at two ends of the culture war raging in many Western countries.

The dominant story for many elite institutions is the idea that we live in a system in which inequality, racism, sexism, and other forms of oppression are all-encompassing and almost impossible to escape. Ibram X. Kendi, author of *How to Be an Antiracist*, captures this worldview when he writes:

> *Our world is suffering from metastatic cancer. Stage 4. Racism has spread to nearly every part of the body politic, intersecting with bigotry of all kinds, justifying all kinds of inequities by victim blaming; heightening exploitation and misplaced hate; spurring mass shootings, arms races, and demagogues who polarize nations, shutting down essential organs of democracy; and threatening the life of human society with nuclear war and climate change.[53]*

In this model, the single oppressive force of racism explains much of what's wrong with the world. On the other end of the political spectrum, we are seeing a strong backlash to this worldview, and its adoption by systems of power. One example is the rise of the reactionaries, also known as the New Right, or postmodern right. Writing about this phenomenon in *Vanity Fair*, James Pogue explains that:

> *...it is not a part of the conservative movement as most people in America would understand it. It's better described as a tangled set of frameworks for critiquing the systems of power and propaganda that most people reading this probably think of as 'the way the world is.' And one point shapes all of it: It is a project to overthrow the thrust of progress, at least such as liberals understand the word.[54]*

Alex Kaschuta is one of the leading podcasters covering this phenomenon. When I spoke to her, she explained that even though it is a disparate movement with many different viewpoints, what unites everyone is a push-back against what they see as the authoritarian creed of 'wokeness' and how it's been adopted by elite institutions like universities, the media, and tech companies.[55] The most well-known reactionary thinker is a blogger and programmer named Curtis Yarvin. His idea of the 'cathedral' is perhaps one of the key tenets of this group:

> *The cathedral is just a short way to say 'journalism plus academia' – in other words, the intellectual institutions at the center of modern society, just as the Church was the intellectual institution at the center of medieval society... all the modern world's legitimate and prestigious intellectual institutions, even though they have no central organizational connection, behave in many ways as if they were a single organizational structure... it always agrees with itself. Still more puzzlingly, its doctrine is not static; it evolves; this doctrine has a predictable direction of evolution, and the whole structure moves together.[56]*

For both Yarvin and Kendi, the establishment has been captured by an all-encompassing form of oppression: an oppression that surrounds us, affects our thoughts, and traps us. And for both of them, even though they're at opposite ends of the political spectrum, the solution is to deconstruct it; to dismantle it from the very core and build something new.

Knowingly or not, both are reviving an ancient myth – one that is both deeply psychedelic in its outlook and profoundly woven into our mythologies around the world. By looking back to this myth and listening to what it has to say, we may find the seeds of a new one for the times ahead.

Gnostic Times

To find that myth, we have to jump back in time twice: first 70 years, then 2,000. The first jump takes us to the 1950s, when two Egyptian brothers discovered an earthenware pot filled with ancient manuscripts while digging for fertilizer. It would prove to be a monumental discovery: a trove of early Christian writings. For the first time, the world had access to the Nag Hammadi manuscripts, also known as the Gnostic Gospels. The Gnostics were a group of early Christians who were persecuted for putting forth a radically different cosmology from that of orthodox Christianity. Until the scrolls were unearthed, all we knew about them came from early church fathers like Irenaeus, who wrote critiques about how awful they were. The ancient Gnostics weren't Christian in the way we might think of now; they had roots both in Hellenistic mystery schools and early Christianity.

Some of the scrolls related a strange creation myth that speaks to the depths of the dynamics we're experiencing today. For the Gnostics, God existed beyond time and matter in an eternal 'fullness' they called the pleroma. While God is singular, it is divided into different aeons, or divine entities that exist within that fullness. As the story goes, at the beginning of time, an aeon called Sophia, a divine feminine wisdom, exists in that fullness with her masculine counterpart. Through an act of passion, she leaves the fullness – and because she does so alone, she accidentally creates a split in the universe. This leads to the world of matter, and a false god: the demiurge.

The demiurge is the one who declares himself the god of all he sees. Bearing a striking similarity to Iain McGilchrist's explanation of the left hemisphere's view of the world, the demiurge declares that he is the god of everything. Sophia tells him he is merely the god of matter, the broken reflection of a deeper divine reality. Some of the scrolls describe him as an abortion: a soulless entity that should have never existed – a god who is blind to true reality, but cannot see his own blindness.

For the Gnostics, humans are trapped in the demiurge's world. He is their version of Moloch. His power is maintained by his army of archons, lifeless entities that perpetuate his world.[57] According to the Gnostics, the archons were immaterial concepts, but also metaphors for power structures. They were the tax collectors, the church leaders, the money lenders.

The Gnostics believed humans are trapped by a false god who infects our minds and deceives us into believing that the world of matter is all there is. But because he isn't truly real, just a shadow of the divine Sophia, no matter how much dominion one has over the physical world, they can only ever create empty imitations of the underlying divine reality that is the true birthright of all people. The harder you try, the more you create a fake, empty, Disneyland reality.

As French theorist Jean Baudrillard described it, such a place is a hyperreality in which it is impossible to connect to anything true or real.[58] For the Gnostics, the jealous god of the Old Testament is the demiurge who tells people not to find their own spiritual wisdom. The serpent in the Garden of Eden is a hero sent by Sophia, the manifestation of divine wisdom, to guide humanity toward enlightenment by exhorting Eve to eat the fruit from the tree of knowledge.

To make sense of this bizarre sci-fi tale told 2,000 years ago, we can turn again to Carl Jung. Jung was fascinated by the Gnostics, perhaps because he saw in their teachings the story of the human psyche writ large. For Jung, we have two key aspects to ourselves: The first is a deep-down, true self, which he called the Self, rooted in the unconscious. The second is an ego, or a personality structure, that exists on top of that, often as a compensation for the shadows we can't access.

When the ego is disconnected from the Self, as the demiurge is disconnected from the divine, it begins to endlessly mimic and replicate deeper truths while never managing to create anything meaningful. Deeper desires become empty expressions of our personality structures. Our capacity for selfless love turns into the quick-fix lust we see on reality-TV dating shows.

Our longing to connect to something bigger than ourselves becomes an obsession with brands or celebrities. The Gnostics and Jung were pointing to a similar kind of insight that we're witnessing as it arises from the edges of culture today – the insight that our division and disconnection are creating a blind, destructive culture that will lead us to ruin. We have fallen into a sociopathic, narcissistic certainty that all that matters is things, and all we need to do is gain power by owning more things. It is the defining story of an identity-obsessed selfie culture.

What binds together the various versions of this myth, which we see across the political spectrum? As different as they may seem at first glance, all of them tell us we're living under an authoritarian system. We are being manipulated by a spell of forgetfulness, and blindness. It is a spell that is slowly stripping vitality from the world. Institutions, people, ideologies –

everything is laid bare and revealed to be fundamentally corrupt, lifeless, and empty.

But the story doesn't end there. The Gnostics believed there was a way out of that spell, a way to liberate ourselves by accessing the divine knowledge of Sophia; to see, through gnosis, divine knowledge, a new reality; to die again and again into the sacred source of all reality. This brings us straight back to the mystical experience elicited by psychedelics.

Sacred Embers

The time has come to pay attention to your hands; to notice the contours of that ember I invited you to hold at the beginning of this journey into darkness. Perhaps the deepest hope we hold is the remembrance that there is something beyond what we can see. This is the divine knowledge the Gnostics saw as our liberation: an awakening into an awareness that we are far greater than our own individuality.

In the beginning of this chapter, I explained Jung's concept of the shadow. The demiurge and Moloch can both be seen as metaphors for the deepest shadows of the human psyche: the blind, destructive, suicidal urges that overtake us when we are profoundly disconnected from anything beyond our own egoic desires. Our blindness to something greater than ourselves can be so deep that even when we have an experience that takes us beyond our egos, we can co-opt and twist it until it serves our own self-centeredness. In the psychedelic world, this is known as 'psychedelic narcissism.'

Despite these potential pitfalls, psychedelics can still help us connect to something beyond ourselves in a real and lasting way. The mystical experience that so many have on psychedelics is an experience of recontextualizing our egos in relation to reality. The experience, as it is identified by researchers, is consistently described as sacred, transcending time and space, impossible to put into words, and filled with a sense of unity and love. As a participant in a Johns Hopkins psilocybin study reported:

> *In my mind's eye, I felt myself instinctively taking on the posture of prayer in my head. I was on my knees, hands clasped in front of me and I bowed to this force. I wasn't scared or threatened in any way. It was more about reverence. I was showing my respect. I was humbled and honored to be in this presence.... It was when I surrendered to this, that I felt like I let go. I was gone... or I should say this earthly part of me was.... I was in the void. This void had a strange and indescribable quality to it in that there was nothing to it but this feeling of unconditional and undying love. It felt like my soul was basking in the feeling of this space.*[59]

What countless spiritual traditions have taught is that the only thing that can bring us beyond our own delusions, existential pain, and ideological fantasies is an encounter with a deeper level of reality. It's the profound, the mysterious, the transcendent, that reframes the very experience of being human and helps us overcome our own blindness. It's the Gnostic connection to true inner wisdom that helps us escape the evils we create through self-delusion. It's waking up from *samsara*, the Hindu description of the cycle of death and rebirth that binds us to this earthly plane, and to suffering. It's leaving the Matrix.

This encounter with the sacred may be, as the Gnostics believed, the best way to escape a rigged game. The sacred, however we may personally experience it, gives us a higher value than profit, or status. It can, if we orient ourselves toward it, help us to overcome the intense incentive structures that so often capture our values.

But having a sacred experience doesn't mean we can renounce the world and stop engaging with it. It isn't a bypass that avoids our pain. Instead, as we saw in the example of Roseman's study, processing collective shadows involves embracing complexity. The demiurge isn't here to be fought, but to be integrated as an aspect of our own psyches within the greater context of the sacred.

Jung said that 'there is no coming to consciousness without pain.'[60] Pain, difficulty, and challenge are core to the psychedelic experience. Psychedelics can help us process and reframe not just our own pain, but the pain of another. They can bring us out of ourselves. When I spoke with Bessel van der Kolk, author of *The Body Keeps the Score: Brain, Mind, and Body in the Healing of Trauma*, he described how he had been a skeptic of the idea of 'vicarious trauma' until his first guided MDMA therapy session. During it, he experienced the pain felt by many of his patients over the years, and how it had impacted him.[61]

This experience of deep empathy was a significant feature in the experiments of Chris Bache, a religious-studies scholar and author of *LSD and the Mind of the Universe* who chronicled 73 controlled LSD trips he took over three decades.[62] For a period of several dosings, Bache experienced what he called the 'ocean of suffering,' or the experience of feeling the pain of thousands of people around the world: the pain of women, of children, and of those affected by war. In addition, he experienced profound unity and connection with all of reality. The two go hand in hand.

It isn't just seasoned psychedelic explorers who have these experiences. My wife, Ashleigh, runs a monthly psychedelic integration group, where people often come to try to make sense of and integrate difficult or meaningful trips. While I was telling her about this chapter, she pointed out that the experience of feeling the suffering of the world comes up in the group regularly. It can be one of the hardest to integrate, but is often experienced as deeply meaningful and important.

The Sacred in Society

The sacred can be related to spiritual experience, but doesn't have to be, and not everyone needs to or should run out and try a psychedelic in the hope of having one. Likewise, we all experience the sacred differently. As anthropologist Nicolas Langlitz has argued, highly materialist scientists who take psychedelics often experience a 'mystical materialism' that enables them to sense a unity between biology, consciousness, and matter.[63]

In fact, on a social level, the sacred doesn't have to be linked to metaphysical changes at all. In some schools of sociology, the sacred is understood to be whatever we set apart as having a higher value than our own self-interest. French sociologist Émile Durkheim argued that society is always in a dynamic between the sacred and profane; that even though modernism has stripped religion away from daily life, we cannot help but make things sacred.[64] For example, in most cultures, human life is seen as sacred. For many of us, our favorite sports team is sacred. Sociologist Jeffrey Alexander has argued that the idea of the sacred moves 'far beyond the realm of traditional, institutional religion to shape political life and civil society more generally.'[65]

A defining feature of the age we're living in is that we make the sacred profane, and the profane sacred. Careening between nihilism and narcissism, a culture that worships perpetual growth has no idea what to do with something that transcends our egos. It becomes a lifestyle choice. We put it on Instagram. We're encouraged to cut the sacred into manageable chunks so that we can microdose our way to success in a fluorescent dystopia. In our desperate, addicted quest for our own salvation, we have forgotten an ancient truth about the sacred. It is recounted beautifully by Peter Kingsley in his biography of Carl Jung, *Catafalque: Carl Jung and the End of Humanity*:

> *Right at the core of Jung's life and experience... lies an awareness that one comes face to face with the reality of the sacred not through sanity, but in the terrifying depths of madness. And there – in the confrontation with madness – is where our normal, collective sanity is seen for the even more horrifying insanity that it really is. Then, every fixed idea one ever had about anything comes permanently crashing down; and the search begins to find some language that can say what everybody thirsts for but almost nobody wants to hear.*[66]

Psychedelics can reliably elicit an encounter with the sacred, but that encounter is not what we expect. What countless shamanic and spiritual traditions teach us is that the sacred is not there to make our lives better. It is not there to confirm our warped ideologies, or for us to post on Instagram for a fleeting hit of dopamine. The sacred is there to shred us. It guides us to the depths of ourselves to find our own delusions and dispel them. Psychedelics were originally called 'psychotomimetics,' because they were thought to temporarily induce insanity. It may be that going temporarily mad is the only way to see how mad we already are.

As Kingsley relates in *Catafalque*, the word 'therapy' comes from the ancient Greek *therapeia*, which means 'caring.' In very ancient Greek history, there was a common phrase, therpaeia theôn, which means 'attending to the divine, caring for the gods and serving them, doing what humans ought to do to make sure the gods are all right.'[67] And this was an important idea, before Plato rejected it and argued that it is the gods who should take care of us. Kingsley views this, and the West's subsequent worship of reason, as a process of disconnecting humanity from a relationship to the divine.

In the context of psychedelics, attending to the gods can be a powerful idea. The sacred isn't there to take care of and look after us. Instead, it is there for us to look after – protecting it from our own blindness; from predatory capitalism and winner-take-all outcomes; from bad incentive structures and all the structural inequalities they bring.

Closing Comments

What I have been exploring in this chapter is what divides us, traps us, and gets in the way of our collective pursuit of a better world. I've argued that psychedelics are not immune from being captured by the worst aspects of our economic systems, and looked at why the question of how psychedelics could change society doesn't often lead us to answers, but to more questions.

At the same time, I've suggested that there is something unique and powerful about psychedelics that can shine through the fragmentation in our hearts and in our societies: their capacity to connect us to something beyond ourselves, regardless of our spiritual beliefs or cultural backgrounds. The psychedelic experience can help us see beyond our conflicts, beyond our status games, and beyond our own limiting beliefs about ourselves and one another.

However, the process we need to undertake to perceive in these new ways isn't simple. It may involve defending the sacred by actively putting boundaries in place against predatory capitalism, and the twisting of revolutionary ideas into corporate-friendly, watered-down versions of what they used to be. It means going to war with Moloch.

As I write this, I notice that I am once more falling into the trap we often fall into when reflecting on darkness and the shadow. I am framing Moloch as something 'out there.' But of course, we are Moloch. We are the demiurge. We are, through our blindness and our collective madness, sowing the seeds of our demise. The greatest hope psychedelics hold for helping us understand ourselves and the world is the manner in which they reconnect us to the sacred. To see why, we have to look at what it actually means to have a 'mystical experience' – what people report about them, and how these experiences could profoundly change the way we understand and experience reality itself.

Chapter Five

A New Reality

Until now, I have been exploring how psychedelics can help us make sense of the complex world we live in. I have looked at how we can apply the mindfulness, openness, and cognitive flexibility that the psychedelic experience unlocks to making sense of culture, the media, and mental health. We've delved into how shamanic discernment can be used to navigate the archetypal landscape of the Internet and give us a new lens on the strange new religions forming online. I have also explored how psychedelic therapy can bring us face to face with our dark sides and integrate them, and why this process of transformation is fraught with peril when it takes place in a system designed to maximize profit and commodify the sacred.

In the final chapters, I ask whether the psychedelic experience can help us evolve; whether the insights and capacities it unlocks can be harnessed toward significant social change. That will bring us to the most transformative aspect of psychedelics: the mystical experience. What is it exactly, and why are scientists, mystics, and shamans all holding or exploring it in different ways? Beyond the therapeutic power of the mystical experience, I explore the social implications of this radical shift in perspective, and why it can play a role in systems change.

Foundations of the Big Crisis

To understand the ramifications of the mystical experience and what it means for the societies we live in, especially as we navigate the Big Crisis, we have to go back to the Age of Enlightenment, between the late 1600s and the early 1800s. It was during this time that the things we take for granted now in the West – individual liberty, reasoned debate, and the scientific method – began to take shape.

Our science, philosophy and conception of reality is still heavily influenced by Enlightenment thinking – specifically, by a French philosopher and mathematician named René Descartes. In the mid-1600s, Descartes sowed the intellectual seeds that would radically change how we view ourselves, our minds and matter. As John Vervaeke explains in *Awakening from the Meaning Crisis*:

> *[Descartes'] whole proposal is that we can render everything into equations, and that if we mathematically manipulate those abstract symbolic propositions, we can compute reality. Descartes saw in that a method for how we could achieve certainty, and he understood the anxiety of his time as being provoked by a lack of certainty and the search for it, and this method of making the mind computational in nature would alleviate the anxiety that was prevalent at the time.[1]*

Just as we are living through the Big Crisis, Descartes was living through a time of great uncertainty and recognized the need for a new paradigm. What he is most famous for is 'Cartesian dualism,' the idea that mind and matter are separate. As a Catholic, he didn't argue that the mind came from matter, simply that the individual mind was separated from matter. Interestingly, he posited that the soul enters the body through the pineal gland, which is the same part of the brain Rick Strassman and others have theorized produces DMT.

Descartes' ideas would have as significant an effect on European society as a huge collective dose of DMT. He began a process that would move past merely separating mind and matter, and toward a worldview that saw *only matter as real*. There is a direct line here to the mystical traditions I explored in Chapter Four, particularly the Gnostic myth that views the human soul as imprisoned by a false god of matter – a god that tells us matter is all there is, who weaves a spell of forgetfulness that prevents us from seeing any reality beyond the things we can touch and measure.

A contemporary of Descartes, Thomas Hobbes, went further and suggested the thinking arose from small mechanical processes happening in the brain. In doing so, Vervaeke points out, he was laying the ground for artificial intelligence:

> ...what Hobbes is doing is killing the human soul! And of course that's going to exacerbate the cultural narcissism, because if we no longer have souls, then finding our uniqueness and our true self, the self that we're going to be true to, becomes extremely paradoxical and problematic. If you don't have a soul, what is it to be true to your true self? And what is it that makes you utterly unique and special from the rest of the purposeless, meaningless cosmos?

I have been exploring different aspects of the disconnect between science and spirituality throughout this book, because psychedelic mainstreaming is brimming with the tension Vervaeke is pointing to. As Erik Davis explained when we spoke, psychedelics sit at the nexus between science and spirituality because they are a molecule (matter) that changes our minds (consciousness).[2]

Four hundred years after Descartes and Hobbes, our scientific methods and outlook are still heavily influenced by their ideas. We still don't really understand the interaction between matter and mind. The most dominant theory about reality is still that matter is the only thing that's real. This is known as 'materialism' or 'physicalism.' Physicalism can tell you what's happening in the brain when you're happy, for example, but it can't tell you what it's like to be *you* when you're happy. Science has no idea what consciousness actually is. This is known as 'the hard problem of consciousness.' From a physicalist viewpoint, your experience is a byproduct of matter. Nothing more than an illusion. A ghost in a meat machine.

In many ways, physicalism can be seen as a necessary step in our intellectual evolution. It played a vital role in allowing us to break free from the constrictive ideas of the Church, and to gain a new view on nature. However, it is also the worldview that many philosophers and scientists now believe is leading us to destruction. It lies right at the heart of the Big Crisis. To see why, we have to recognize what physicalism *can't* tell us about life.

Mystical Experiences

As we've seen, the mystical experiences people have on psychedelics are often experienced as encounters with a reality greater than themselves, or an animating force beyond time and space. Such experiences are often profound encounters with a transpersonal, non-physical reality that is experienced as completely real, and that many describe as one of the most significant experiences of their lives. Many of the most staunchly physicalist psychedelic researchers and clinicians recognize that this experience leads to lasting therapeutic effects.

This presents a paradox, because what actually cures us is an experience that often leads people to question the truth assumptions of the scientists leading the research. This disconnect between our experience of the world, and what our culture tells us is 'real,' which we see in psychedelic science, is a microcosm of a much larger tension in modern technological cultures.

We experience ourselves as alive and conscious, but scientific materialism tells us that our love, our dreams, our deepest experience of the sacred, are simply a byproduct of physical processes.

Psychological research can be quantitative – for example, gathering data through brain scans or by collating questionnaire responses or data from cognitive tasks – or it can be qualitative in its examination of what people say about their experiences. These are distinct ways of knowing (epistemological categories that explain how we know what we know, and the vehicles by which we receive this knowledge, such as reason, sense perception, intuition, inherited information, etc.). I can't come back from a psychedelic experience and make a definitive claim that the entities I saw were real. Likewise, the team at Imperial can't look at the brain images while I'm under DMT and say, 'What we're seeing *is* your experience.' What they are observing are neurological correlates of my experience: what's happening in my brain while I am in an altered state. Saying that they are observing my experience would be like claiming that the wake left by a boat is what's causing the boat to move.

This has been called the 'explanatory gap' between subjective experience and objective data. It's a gap I've wrestled with endlessly as I've practiced Eastern meditation traditions and psychedelic exploration. But it wasn't until my final dosing on the trial that I began to consider different ways in which the gap can be closed, and that if we can do that, we may unlock a completely new way to make sense of the world and radically change our behavior.

The DMT extended-state trial was only supposed to include four dosings. My fourth had been beautiful but not particularly strong. It felt like the completion of a series of lessons around intimacy and emotional truth, and ended with a moment in which the Teaching Presence encouraged me to let go and become intimate and forgiving toward myself. It left me in a space of awe and gratitude toward the DMT experience. However, it hadn't given me what I was looking for in my 'experiment within the experiment': no metaphysical insights to take back into the world and wrestle with; no 'aha' moments that changed my view on the nature of reality.

But that all changed in the spring of 2022. Imperial got in touch and invited those of us on the trial to an extra dosing. I would find out later that they'd worked out the dosing based on our experiences and wanted to offer us what they believed was a perfected dosing experience.

It was the highest dose yet, and it came on hard and fast. I almost vomited, and as the scientists and medics moved around me to get a bucket and Lisa held my hand, the Teaching Presence told me I had to believe in myself. After much resistance, I made a mental statement that I did, and an instant later, a portal ripped open ahead of me.

Within that portal, exquisitely organic shapes overlapped, reaching out to caress themselves. I focused my attention and flew through it, transitioning seamlessly and undramatically into another universe. It was teeming with intelligence and life. I couldn't see all of it, but I could feel it around me, mysterious and unknowable and all very *busy*.

A huge surge of happy energy emerged from the cacophony, and dozens of entities rushed toward me. They looked like chinchillas – semi-translucent, made of shifting lines, darting back and forth and very excited to see me. All at once, they started communicating in complex symbols that warped the space around us. I had absolutely no idea what they were trying to express, but their energy was infectious and I couldn't help but smile. I tried to send them a feeling, an intention of 'good to meet you,' but it had zero effect.

They kept chattering. The space around us was dripping with intricate symbols and I was half stunned by the otherness of it all but tried to mimic what they were doing. I paid attention to the symbols and started generating my own. It's hard to describe what this involved, but essentially, I imagined them and consciously projected them out of my body. This sent the chinchillas into a frenzy of excitement. They started shooting in and out of my body, trying to show me something.

They opened another portal. Instinctively, I knew it led to where they lived, and that they wanted me to follow. For a moment, I was tempted, but something held me back. I had done a session with Trish (the coach I worked with to prepare for and integrate my experiences) before the dosing during which she had encouraged me to check in with myself and keep my own sovereignty when communicating with entities. In that moment, I had the feeling that there was something important in this larger space, so I told them in my own words that, while I was grateful, I was going to stay here. They seemed to accept this and very quickly appeared to get bored.

I noticed a new shape close by. It looked like a black squid. The boundaries between everything were unclear, and I couldn't tell if this squid was one of the chinchillas or something different. After a few moments of observing it, I realized it was distinctly different. It was also less interested in me, and overall, it had a slower, more reserved energy.

I started to wonder whether this was a different species, and whether it occupied a different ecological niche here, wherever 'here' was.

The Teaching Presence burst into my awareness from nowhere.

'That's right,' it said. 'That is the right way to conceive of what's going on. This is an ecosystem with thousands of intelligences in different niches. And humans exist here, too.'

This floored me. I looked around the huge universe I had entered and had a flash of understanding that human consciousness extends outward to this space in some way. As I looked around, I noticed more of the complexity, and some hidden order within it. Everything here existed in a niche. This ecosystem of consciousness was as competitive and cooperative as any we might find on Earth.

I asked the Teaching Presence why, if there was a whole ecosystem of intelligences, we only encounter some entities and not others. It explained that what you encounter depends on your intention, and also on random chance. Intention acts as a kind of organizing principle, attracting particular entities with a matching signature. But taking DMT is also similar to parachuting into the middle of the Amazon rainforest, so what I was now perceiving was just whoever happened to be in the part of the ecosystem where I landed. The Teaching Presence went on to say that there were countless layers of the ecosystem. This was only one, and it lay at the foothills of something much vaster.

In a flash, I was shown a glimpse of the true scale of the ecosystem of consciousness. It is hard to describe just how awed I felt. It was vast beyond anything I have ever witnessed. I knew myself as just one blip of consciousness among countless trillions. I felt absolutely tiny, but not insignificant. It was humbling, sublime, sucking any sense of meaning-making or reason away from me. All I could do was surrender to the undeniable reality of my own minuscule perspective within the scale of a thriving universe teeming with life.

The Overview Effect

What I was experiencing is just one example of a mystical experience that we can have while on psychedelics. The closest correlate to the kind of experience I was having is the 'overview effect' reported by some astronauts viewing Earth from space for the first time. As astronaut Edgar Mitchell told *People* magazine in 1974: 'You develop an instant global consciousness, a people orientation, an intense dissatisfaction with the state of the world, and a compulsion to do something about it. From out there on the moon, international politics look so petty. You want to grab a politician by the scruff of the neck and drag him a quarter of a million miles out and say, "Look at that, you son of a bitch."'[3]

There are many varieties of mystical experience, which is what led the team at Imperial to develop a 'metaphysical beliefs questionnaire' that we filled out after each dosing. However, what these experiences share is that they connect us to something beyond ourselves, to a transpersonal perspective.

In my case, I had the experience of my own personal consciousness in relation to that of many other conscious entities. I didn't travel in a rocket to see Earth from space, but as many people report from DMT and other sacred experiences, like near death, I am convinced that what I encountered was as real as a space shuttle. It stayed with me for months. In that time, I would often sit on the roof of our houseboat and look out at the canal with a sense of vastness around me. I would absorb the bustle of London, with all its many sentient people, and feel we were just the tip of the iceberg with respect to what's really out there.

It was also very grounding, and made me feel what is known in the 12 Step philosophy as 'right-sized ego.' The experience was fascinating in that it allowed for hyper-individualism and hyper-connection at the same time. There was no need for the ego-bashing that we see in some Western interpretations of Eastern philosophy, because no matter how big my ego was (and it can get pretty big), it would always be tiny and insignificant within the true vastness of the universe. In turn, an awareness of the vastness of the universe made ego inflation feel absurd.

The only way to account for this type of experience through the lens of our existing science is that it is generated exclusively in the brain. The reason I was wearing an EEG helmet was so that the researchers could get readings of measurable, quantifiable electrical phenomena in my brain. However, none of this accounted for my change in perception. You could see changes in the electrical activity in my brain, then infer from this how DMT interacts with neurochemistry, and how it maps against what people report they saw and felt. This is important data, but scientists aren't actually measuring the psychedelic experience itself. They are measuring the neurological correlates of a subjective experience.

This isn't to suggest that there is no link between subjective experience and brain chemistry, simply that we don't understand what that link is. I couldn't take my insights in my hands and show them to anyone. The only way the scientists could learn about them was from me recounting my subjective experience or filling out a questionnaire, which they would then interpret through their own subjective experience and compare against blood and EEG data. It's an important process, but it is by no means the most interesting or useful way to understand how psychedelics change our perspective.

Physicalist Neuroscience

We can find more interesting potentials for psychedelic changes in consciousness if we look more deeply at the contradictions of a physicalist perspective on consciousness. If you think the eyes you're using to observe nature, your own consciousness, are simply a byproduct of matter, then you come to a philosophical dead end, as that very consciousness is the only thing you can be sure is real. As Andrei Linde, renowned physicist and pioneer of the inflationary universe theory, points out:

> Let us remember that our knowledge of the world begins not with matter but with perceptions. I know for sure that my pain exists, my 'green' exists, and my 'sweet' exists... everything else is a theory. Later we find out that our perceptions obey some laws, which can be most conveniently formulated if we assume that there is some underlying reality beyond our perceptions. This model of the material world obeying laws of physics is so successful that soon we forget about our starting point and say that matter is the only reality, and perceptions are only helpful for its description.[4]

While many neuroscientists recognize the importance of qualitative data, or information about people's subjective experience, data that can be quantified is perceived as far more valuable. This leads to a strange situation in which thousands of people are reporting that they have encountered entities and worlds they experience as undeniably 'outside of me' as a chair in a room, but the scientists studying them are often using a model that can't accept that such things could be real.

It's interesting to inquire into whether psychedelic neuroscientists themselves actually behave as though their own conscious experience isn't real. This was the question that inspired anthropologist Nicolas Langlitz to make psychedelic neuroscientists the subject of a research project, which he published in his book *Neuropsychedelia: The Revival of Hallucinogen Research since the Decade of the Brain*. He spent time in the lab of psychedelic researcher Franz X. Vollenweider, where he interviewed dozens of psychedelic scientists to get a sense of how they interpret the psychedelic experience, and how their beliefs about consciousness influence the way they live their own lives.

Langlitz points out that almost all of the researchers he observed were physicalists who believed that their experience of the world was caused by the physical processes in their brains – and often, that they themselves, as well as their concept of a 'me,' weren't actually real. As Langlitz explains: 'When talking about themselves, the hallucinogen researchers I worked with frequently referred to brain anatomy and chemistry.... Things said in a fit of temper were attributed to malfunctioning frontal lobes. One scientist who had just fallen in love was joking about his oxytocin level. E-mails were signed "Serotonergically yours."'[5]

But when life got real or they were talking about something emotionally meaningful, 'the neuroscientists readily reverted to talking about themselves rather than their brains.' Langlitz is revealing a performative contradiction here, one explained well by philosopher and computer scientist Bernardo Kastrup:

> *Try as we might, we don't experientially identify with neurophysiology; not even our own. As far as our conscious life is concerned, the neurophysiological activity in our brain is merely an abstraction. All we are directly and concretely acquainted with are our fears, desires, inclinations, etc., as experienced – that is, our felt volitional states. So, we identify with these, not with networks of firing neurons inside our skull.*[6]

It may seem relatively minor, but this huge gap between what science tells us is going on with consciousness and what we're experiencing leads straight to the most important concept in this book. Why is it that for me and so many others, it just doesn't feel true that consciousness is a byproduct of neurons? Perhaps because it isn't.

A Conscious World

What if, instead of the dead machine we have been told it is, the world is actually alive? What if it isn't just 'things' that are real, but the consciousness that perceives those things, as well? The theory that reality is fundamentally a mental process rather than a physical thing is known as 'philosophical idealism.' Perhaps the best-known idealist is Bernardo Kastrup, who was quoted earlier. Kastrup holds a PhD in computer engineering and has worked in CERN, also known as the European Organization for Nuclear Research. He also holds a PhD in philosophy, and has been challenging physicalists on what he sees as the fundamental flaws in their view of the world.

When we spoke, he told me that he's noticed a significant shift taking place in the last few years in debates on the nature of reality, and he's found it hard to find anyone who will debate from a staunchly physicalist position. This is partly due to what we're discovering in quantum physics, which I will explore later in the chapter. But it is mainly, in Kastrup's view, because the physicalist position doesn't make any sense.[7]

The idea that consciousness is the foundation of everything can seem counterintuitive. It wasn't always so, and idealism has a rich philosophical tradition around the world, appearing again and again in Eastern and Western philosophies, as well as animistic cultures around the world. More recently, it has arisen as a theory to explain the strangeness of quantum mechanics, which is the physics of atoms and subatomic particles.

What quantum mechanics reveals is that the seemingly solid world around us is, when you really look at it closely, turns out not to be solid at all. Classical physics is excellent at helping us predict and understand how things larger than atoms behave, such as how an apple will fall from a tree. But when scientists began looking deeper to see what was going on in the interaction of atoms and smaller particles, classical mechanics went out the window.

As well as realizing that matter is wave-like in nature, flowing like water rather than fixed like a rock, scientists were stunned to discover that the very act of observing tiny subatomic particles changed them. The smallest parts of matter that we know of exist in a field of possibility, or what's called 'superposition.'

The act of someone observing and measuring these particles in an experiment collapses that possibility field, and the particle goes from being either a one or a zero to becoming 'real' in one sense of the word; the field of possibility collapses and the particle falls into its final 'state.' It's hotly debated whether that 'measurement' needs to be made by a conscious human being: It could be that a fly buzzing its wings or a rock falling down a mountain is nature 'observing itself' and collapsing the possibility field around it.

This can be hard to grasp, but a useful metaphor can be found in your own mind. If you were to stop reading now and to imagine three possible futures for yourself, you might notice that they seem hazy and uncertain. If you made a series of choices that led you toward one of those futures, you would be collapsing all three possibilities into that future and you would have a lived experience of it as more 'real' and 'solid.'

Kastrup and physicists Henry Stapp and Menas C. Kafatos argue in an article in *Scientific American* that what we're learning from quantum mechanics points to a reality that is fundamentally mental rather than physical:

> *Our argument for a mental world does not entail or imply that the world is merely one's own personal hallucination or act of imagination. Our view is entirely naturalistic: the mind that underlies the world is a transpersonal mind behaving according to natural laws. It comprises but far transcends any individual psyche.... The claim is thus that the dynamics of all inanimate matter in the universe correspond to transpersonal mentation, just as an individual's brain activity – which is also made of matter – corresponds to personal mentation.*[8]

Elsewhere, Kastrup argues that a view that perceives the act of seeing as primary is philosophically sounder than one that assumes matter is the only thing that's real:

Consciousness is the only carrier of reality anyone can ever know for sure; it is the one undeniable, empirical fact of existence... underlying all reality is a stream of subjectivity that I metaphorically describe as a stream of water (water being analogous to consciousness). Inanimate objects are ripples in the stream, experienced subjectively by the mind-at-large that is the stream itself. Living creatures are localizations of the flow of water in the stream: whirlpools. ... The body–brain system doesn't generate consciousness for exactly the same reason that a whirlpool doesn't generate water. And since there is nothing to a stream full of ripples and whirlpools but water in movement, all reality is simply consciousness in movement.[9]

If the universe is one vast consciousness which we're all a part of, this opens up the possibility that we can in fact access different aspects of that consciousness through DMT experiences. It opens the possibility that the conscious intelligences we meet in these encounters are as real as anything we can touch. If we take this seriously, then alien intelligence no longer has to be located 'out there' in the depths of space, but may be 'non-local' and exist, like the particles that make up your eyes as you read this, in a field of possibility that transcends space and time.

What I experienced during my final dosing only made sense to me when seen through this lens. Finding myself in a vast ecosystem of intelligences, I felt a true sense of awe and a deep curiosity. Any attempt to try and boil down what I was experiencing to 'my brain on drugs' would have felt at odds with my empirical observations.

After the Teaching Presence told me that I was in an ecosystem of intelligences, I had the sense that it included intelligences of varying levels of complexity, with different goals, cultures, values, and who knows what else. I also felt intuitively that it would somehow be possible to access different parts of the ecosystem with the right know-how. During this time, the recording through the headphones periodically asked me, 'Entities: Yes or no?', to which I could only give a halting 'yes.' The question started to seem absurd, like being asked, 'Life: Yes or no?' while standing in the middle of a rainforest.

I asked the Teaching Presence, 'If humans wanted to be in contact with these other intelligences through this technology, could we? Is there a gatekeeper?'

'There is no gatekeeper,' it answered. 'This is not a hierarchy in the way you understand it, or an ecosystem in the way you understand it.'

'So who decides who gets to be part of it? How would we travel here and make contact?'

'You all decide,' it said. 'Every species is the gatekeeper of its own entrance.'

Throughout this dialogue, I started to notice that something about the Teaching Presence had changed. It was quite subtle, but it felt like I was talking to more than one entity.

'Who am I talking to?' I asked.

Everything shifted around me. I dropped into a box-like vortex. The Teaching Presence took physical form in front of me, showing up as a series of blocks stacking atop themselves in a strange and playful maze. I worried that maybe it was stumped for words, and maybe I had been talking to myself – and now, I'd asked a question that my own mind couldn't answer. Finally, it spoke.

'I am you but not you, and we but not we.'

The ambiguity was strange: I was talking to myself, but I also wasn't. I felt that I was talking to multiple intelligences at the same time. I didn't dwell on this paradox, because I realized I probably only had a few minutes left until the dose wore off – a few minutes to ask the questions I really wanted answers to.

Panpsychism

Perhaps the biggest question any of us can ask is, 'What is reality?' As we have seen, physicalist science argues that reality is matter: the things you can touch, feel, and measure that exist in space. On the other end of the spectrum, we have idealists like Kastrup who argue the flip side: The only thing that's real is mind.

It can be hard to grasp the concept that the things you are seeing right now, and that you can touch and feel, aren't ultimately separate physical objects. It can also be hard to grasp the concept that your experience of consciousness and free will is just an illusion.

These are not the only views on consciousness. A popular view among people who take psychedelics is panpsychism. Panpsychism is similar to idealism in that both positions see consciousness as a fundamental quality of the universe. But there are subtle differences, as philosophy professor and panpsychist Philip Goff explains:

> The main difference is that whilst panpsychists think that the physical world is fundamental, idealists think that there is a more fundamental reality underlying the physical world. How can a panpsychist think both that the physical world is fundamental and that consciousness is fundamental? The answer is that we believe that fundamental physical properties are forms of consciousness.[10]

Perhaps the most famous panpsychist is the 17th-century philosopher Baruch Spinoza, who argued that matter and mind are 'of one substance.' Peter Sjöstedt-Hughes is a philosopher and author of *Philosophy and Psychedelics: Frameworks for Exceptional Experience*. In it, he points out that Spinoza's response to Descartes' dualism was that 'mind and matter are not two substances that mysteriously interact but are rather two (of an infinity of) expressions of the same substance – a substance he names God or Nature.'[11]

When I spoke to Iain McGilchrist, he shared that he views matter as 'a phase of consciousness' in a similar way to how ice is a phase of water – a view that struck me as a useful metaphor for panpsychism.[12] In a panpsychist view, matter and consciousness are one. Every part of the universe contains consciousness.

Philosopher Matthew Segall has argued that psychedelics can bring us into a lived experience of this view, toward what he calls a 'psychedelic realism.' Drawing on the work of philosopher Alfred North Whitehead, who is an influence on Sjöstedt-Hughes and McGilchrist, he suggests:

> *The value of psychedelics for philosophy is precisely that the mind-altering, boundary-dissolving, world-enchanting experiences they precipitate force the issue. Consciousness reveals itself to be less like the on/off switch for a ghost-like observer hidden somewhere inside the skull. And more like a transcranial kaleidoscope with a variety of experiential modalities, each revealing a new facet of reality.*[13]

Drawing on Whitehead's version of panpsychism, Segall explains what this shift entails:

> *Rather than the thinking ego... remaining aloof from the world and alienated from its own body, Whitehead reimagines our consciousness as part of the same reality it is attempting to know. My thinking arises out of and perishes back into a cosmic network of creative events. Whitehead invites us to step out of Cartesian solipsism into a panpsychic cosmic process wherein everything becomes a kind of thinking thing.*[14]

What this would mean is that it's not just you, your friends, and the trees outside who are conscious. Everything is. That doesn't mean the chair you're sitting on is questioning the life choices that led it to this point, but that at some level, every atom that comprises it is conscious.

This idea isn't limited to philosophy. A number of prominent scientists support the validity of a panpsychist view, including Sir Roger Penrose, David Chalmers, and Christof Koch. This shift in perspective doesn't mean we do away with science. Instead, it brings science into a completely new paradigm.

Why Does It Matter?

So, why does it matter how we view matter? Because it could fundamentally change how we think and act. For the rest of this chapter, I make the case as to why the capacity for psychedelics to change our view on consciousness is their single most important contribution to improving how we make sense of the world.

Psychedelics give us a lived experience of a conscious universe. In doing so, they often permanently change our metaphysical beliefs. In a study published in *Nature* in 2021, Chris Timmermann et al surveyed more than 800 psychedelic users using the same 'Metaphysical Beliefs Questionnaire' that we filled out during the DMT study. They found that 'Psychedelic use is associated with shifts in metaphysical beliefs away from hard physicalism or materialism. Attending a psychedelic ceremony was associated with shifts away from hard-materialistic views, and items associated with transcendentalism, non-naturalism, panpsychism, primacy of other realms, dualism and solipsism/ idealism, with some changes enduring up to 6 months.'[15]

This shift is highly significant. The fragmentation I have explored throughout this book owes a lot to the cultural story that we are meaningless blips in a cold, dead, uncaring universe. A shift to a perspective with consciousness at its core can radically change the foundations of our behavior. Perhaps more than anything else, it can transform our relationship to the rest of nature. As Sjöstedt-Hughes shared with me:

> *I see that the hard problem of consciousness and the ecological crisis have the same origin. The origin is this: You suddenly look at nature as a dead machine. That was thanks to Descartes, Galileo, Francis Bacon, and others. If you see the world as a dead machine, as quantifiable and not qualitative, that means you can't get mind out of it. This is a hard problem of consciousness. How do you get mind from meat? You also get an ecological crisis. You think nature in itself has no intrinsic worth. The only thing with intrinsic worth is humans. This is a Cartesian point of view. And so, we might as well exploit the world for ourselves and for beings that have sentience, because in itself it has no value. This led to the Industrial Revolution, and to the ecological crisis.*[16]

The profound importance of this shift in perspective on a collective scale is hard to overstate. In our existing cultural story and the scientific metaphysics that underpin it, we are oriented toward *things*. Only things are real, so accumulating them becomes the only sensible thing to do. We may as well focus on perpetual economic growth, working endlessly on our own self-fulfillment and consuming our limited resources to have more fun – because there is no point in doing anything else. Nothing matters, so why should we worry about our impact on the planet? You, everyone you love, your dog or cat and all the other animals and plants are ultimately just dead meat that's temporarily animated. Nothing is sacred, because this story also tells us that science will eventually remove all the mystery from the world.

If, instead, we open our eyes to the possibility that *the world is alive* and we are all part of a larger consciousness, we begin to orient toward the *quality* of lived experience – not just our own, but everybody's. We see ourselves fundamentally as an aspect of something greater, a part of a vast intelligence so far beyond us that we can only marvel at its mystery.

Our suffering becomes shared, as so many people spontaneously experience during psychedelic journeys. Our hopes and dreams weave together. Our strife and disagreements are qualities to be experienced rather than threats to be extinguished. And when we look out into the cosmos, we don't feel alone and scared, but tiny and beautiful.

When I asked Bernardo Kastrup what he thought the implications would be of a widespread adoption of an idealist perspective, he answered:

> *It changes everything. Our sense of plausibility is culturally enforced, and our internalized true belief about what's going on is culturally enforced.... They calibrate our sense of meaning, our sense of worth, our sense of self, our sense of empathy. It calibrates everything.* Everything. *So, culture is calibrating all that, and it leads to problems if that cultural story is a dysfunctional one. If the dysfunctional story is the truth, then, OK, we bite the bullet. The problem is that we now have a dysfunctional story that is* obviously not true.... *If you are an idealist, you recognize that what is worthwhile to collect in life is not things. It's insights.... Consumerism is an addictive pattern of behavior that tries to compensate for the lack of meaning enforced by our*

current cultural narrative.... We engage in patterns of addictive behavior, which are always destructive to the planet and to ourselves, in order to compensate for that culturally induced, dysfunctional notion of what's going on.[17]

A conscious world is an ancient idea. We find it in the Hindu concept of *tat tvam asi*, or 'thou art that,' the concept that there is no separation between you and reality. It was this concept that partly inspired Robert Pirsig to write the best-selling philosophy book of all time, *Zen and the Art of Motorcycle Maintenance*. In this and later books, Pirsig developed a 'Metaphysics of Quality' in which he argues that the experience of existing is the foundational truth of reality, and that quality itself is the fundamental force of the universe. He points out that this brings power back to ways of knowing that are shut out by a physicalist metaphysics. As he writes in his book *Lila: An Inquiry into Morals*:

The Metaphysics of Quality subscribes to what is called empiricism. It claims that all legitimate human knowledge arises from the senses or by thinking [based on] what the senses provide. Most empiricists deny the validity of any knowledge gained through imagination, authority, tradition, or purely theoretical reasoning. They regard fields such as art, morality, religion, and metaphysics as unverifiable. The Metaphysics of Quality varies from this by saying that the values of art and morality and even religious mysticism are verifiable and that in the past have been excluded for metaphysical reasons, not empirical reasons. They have been excluded because of the metaphysical assumption that all the universe is composed of subjects and objects and anything that can't be classified as a subject or an object isn't real. There is no empirical evidence for this assumption at all.[18]

Zen and the Art of Motorcycle Maintenance might be as popular as it is because orienting toward quality means orienting toward the truth of our own lived experience. Speaking the truth of what's going on for us in the moment is the idea behind the practice of inquiry, and of therapy. It is the secret sauce of transformation. It deepens our relationship to reality as it is.

If orienting to the truth of our lived experience is transformative on a personal level, it stands to reason that a collective orientation in the same way could be transformative on a social level. The Whiteheadian philosopher Charles Hartshorne has argued that we're stuck with physicalism only because we lack the imagination to conceive of something else, what he calls the 'prosaic fallacy.'[19]

Segall and Sjöstedt-Hughes both argue that psychedelics may be exactly what we need to overcome this. As Ashleigh said while I was talking to her about this chapter, 'Psychedelics make the implausible plausible.' They invite us to imagine something new. They invite us to revolution – not a revolution to bring about a new ideology we force onto others, but instead, a revolution in our understanding of who we are.

From Game A to Game C

At its core, the Big Crisis is a crisis of metaphysics, a dysfunction at the very deepest level of who we think we are, why we're here, and what is worth living for. A cultural story based on the idea that there is no meaning to existence or the universe will never birth a new worldview that allows us to thrive in harmony with the planet and one another.

Right now, there are millions of people working to try and change the world. They work in systems change, activism, environmental NGOs, innovative startups, or improving their communities, workplaces, and schools. The vast majority of us want to create a fairer, more sustainable world. One of the key questions that arises is: At what level should we enact change?

Should we try to change the values of our workplace to be more inclusive, or do we need to descend a level deeper – to the values of society? A well-known phrase in business is 'culture eats strategy for breakfast.'[20] It's often falsely attributed to management consultant and writer Peter Drucker, but its origins are unknown. Its enduring popularity may speak to an important truth: The value system we are part of is far more impactful than the strategies we put into place. For example, a shared culture that empowers the people within it is more important than a strategic plan that is meticulously designed. Our strategies for changing the world are inspired in part by our particular cultural values.

I propose that metaphysics eats culture for breakfast. What we believe to be real is the most significant factor in our culture, which influences our thoughts and emotions, which change our values, which influence our institutions and political policies. The change has to happen at the deepest level if it's going to have any significant impact.

One particular systems-change community I've been involved with is called Game B. A loose collection of different thinkers, it's a decentralized community founded on the idea that if the cultural and economic system I described through the Moloch metaphor is Game A, we must consider what it means to play a new game, which we can call Game B.[21]

Game A is based on winner-take-all values that inevitably lead to significant wealth disparity, alienation, environmental collapse, and war. Game B seeks to move toward a world built on a philosophy of win-win – a more collaborative, generative, compassionate system that promotes genuine expression, connection, and purpose over endless consumption.

But where will we find these values? What will they be based on? I don't believe they can be drawn from the same soil as our existing problems. The mystical experience gives us insights that come from far beyond our current conceptions of ourselves, and reality itself. In the same way, it may be that effective systems change can only come from outside of our existing frame on reality.

That is what the sacred is: the mystery beyond our current conceptions. It's what lies outside of the game itself; the sacred transforms this game the moment we become aware of its existence. What the psychedelic experience reveals to many is that outside of our own frame is a sacred unity – an aliveness, a purpose, an intensity of being that we have no choice but to surrender to.

If Game B is an analogy for trying to change the social game we're playing, then Game C would be a change that involves the recognition that any significant systems change must come from a re-conception of the role of consciousness in reality; a shift in how we view ourselves at the deepest level; an orientation toward the sacredness of our conscious experience that can be the soil from which we draw our values, ideas, and possibilities.

To see how this could actually change our collective behavior, we can revisit the example of the fish farm from Chapter Four. In the example, the key driver of everyone's behavior was economic survival. The value that was being selected by the farmers around the lake was profit and consumption, because our cultural concept of value is based on economic value.

As Daniel Schmachtenberger has said, 'If a dead whale is worth a million dollars on a fishing boat, and a live whale is worth nothing, that's a value system... that then incentivizes behavior, but it also incentivizes psychopathy. I have to deaden to be incentivized to do the thing that is incentivized by the system, or somebody else does and I am just not effective in the system.'[22]

Just as there is a gulf between what science tells us consciousness is and what we experience, there is a tremendous gulf between what our system tells us to do and what we often want to do. C. Thi Nguyen calls this 'value capture.'[23] Another way to put it is that we can only thrive in Moloch's world if we kill our own souls.

If, instead of valuing the quantity of fish that can be farmed, the system incentivized the quality of experience for the fish, the lake, and the farmers, the outcome of that thought experiment would be radically different. However, we are currently in a situation in which we have to find a way to hold on to our souls while the system tries to take them from us. Systems theorist Jim Rutt, one of the founders of the Game B movement, has often spoken about the necessity of Game B being able to outplay Game A, so that Game B can be perceived as a more attractive and successful way of life. In order for that to happen, it has to give us what Game A never can: meaning, purpose, aliveness, growth, and soulfulness.

Closing Comments

Throughout this chapter, I have explored the deeper philosophical assumptions held not just by science, but by the richest societies in the world. I have looked at how these assumptions drive the Big Crisis, and at the complex ways in which the mystical experiences we have on psychedelics can bring us to a new perspective about the nature of reality. Most importantly, I've suggested that the mystical experiences we access on psychedelics can inspire us to reevaluate the belief that the universe is a dead place devoid of meaning. Instead, it can open us to experiencing the world around us as vibrant and alive, full of mysterious purpose and wonder.

Returning to this sense of wonder is essential for our survival as a species, and access to this kind of experience is needed now, more than ever. In developed nations, we have inherited a culture of acquisition, exploitation, and disconnection from the rest of nature. An increasing number of people are desperately trying to find alternative ways of being and seeing that can help us to play a different kind of social game – before there's no game left to play.

An encounter with the sacred is essential in helping us get there. It gives us an experience of a higher value than maximizing quantity, and wakes us up to the quality of our lived experience, and that of other animals and the plants that sustain us. And there is no practice, substance, process, or experience we know of that can do that faster and more reliably than psychedelics when used with great care and preparation. But how to get to that world from the one we live in seems like a colossal leap. And so, with only a few minutes left in my dosing, I tried to ask as many questions as I could.

I told the Teaching Presence, "A lot of us feel an urgency around fixing our problems as a species, because we're running out of time. You don't seem to share that urgency."

"In the big picture, it doesn't matter if you go extinct," the Teaching Presence replied. "It only matters to you."

There was a matter-of-factness to this statement, and it was communicated without any malice. It was a moment that would stay with me for some time. As that statement churned in my unconscious and I slowly integrated it in the months to follow, I would reflect on what is one of the most profound and important processes we can go through in a psychedelic experience: the cycle of death and rebirth. This fundamental process is all around us in the natural world, in our own psychological transformations, and in many approaches to systems change. For the new to be born, the old has to die.

My final dosing wasn't over, and I had only a short time to keep exploring. What I wanted to explore was the idea of transformation itself – the question of what it would mean for us to bring about a revolution that is not destructive, but generative. A revolution forged not by fire, but by a reverence for something beyond ourselves.

Chapter Six

A Reverential Revolution

True revolution is born not from changing what we do, but changing how we see ourselves. Widening the frame of what we understand ourselves to be at the very deepest level of existence is a process not just of changing our own personal or even cultural cognition, but of changing how we *relate* – how we relate to one another as individuals, as well as how our whole species relates to the wider ecosystem of the planet and the cosmos. Billions of distorted and broken forms of relating lie at the heart of the Big Crisis, and the hope for transformation lies in bringing those pieces back together.

One thing psychedelics teach us is that this change can begin with how we relate to the natural world. As we saw in Chapter Three, the Yaminahua, Asháninka, and Shipibo-Conibo, who use psychedelics ceremonially, have this aliveness built into their cultural story, despite being thousands of miles apart. The rocks, trees, even bubbles in a pot of ayahuasca are populated by conscious entities with personhood. While this precise view isn't necessarily one we would realistically adopt in the West, it speaks to an alternative relationship we can have to the environment, one that allows us to see it as alive and worthy of respect.

Pioneering anthropologist Luis Eduardo Luna has been studying ayahuasca traditions in the Amazon for more than 40 years, and has argued, 'We must rescue animism…. Animism is not a philosophy, it's not a religion; it's a way to relate to the nonhuman world. We need to change our relationship with the nonhuman world. And it has to be a relationship – a two-way subjectivity, or intersubjectivity.'[1]

A 2022 study by Johns Hopkins investigated how psychedelics change the way people attribute consciousness.[2] According to the paper's press release, 'The study found that among people who have had a single psychedelic experience that altered their beliefs in some way, there were large increases in attribution of consciousness to a range of animate and inanimate things. For example, from before to after the experience, attribution of consciousness to insects grew from 33% to 57%, to fungi from 21% to 56%, to plants from 26% to 61%, to inanimate natural objects from 8% to 26% and to inanimate manmade objects from 3% to 15%.'[3]

When I spoke to Michael Pollan about this study, he was positive about the results, but also pointed out that we have to be careful to recognize that the kinds of people who take psychedelics may also be the kinds of people who are already more likely to have a more panpsychist perspective. Thus, further research is needed to figure out if psychedelics can have this effect on different populations.[4]

It's also worth noting that even if psychedelics can change the majority of people's perceptions of reality, this isn't enough by itself. As Luna points out in response to the increased number of people taking psychedelics and opening up to a more animist perspective, 'What matters is what you do with that experience. What is your vision, your relationship with your family, with your neighbors, with your society, with your garden, with the world? If it has changed, great; if not, it's all just narcissistic, like going to a very good film. A very special film, but that's all.'[5]

One way to ensure that these animistic awakenings lead to real systems change is to combine psychedelic insights with biology and policy. As mycologist Paul Stamets has argued, fungi can give us a blueprint for a much deeper pattern in nature – a pattern of interconnected, supportive networks. He sees mycelium, the root network of fungi, as 'an exposed sentient membrane, aware and responsive to changes in its environment.'[6] We don't even have to stretch our current physicalist definitions of consciousness very far to recognize that fungi and plants are conscious.

And if we change our conception of consciousness so that non-human consciousness is as important as human consciousness, we can begin to change our legal system to match this view. There is a growing movement to give personhood to non-human entities, like rivers and mountains, which would allow them the same legal rights as a human being. In 2017, New Zealand passed a law granting personhood to the Whanganui River. As reported by the *Independent*: 'The law declares that the river is a living whole, from the mountains to the sea, incorporating all its physical and metaphysical elements.' The change in law recognized the Māori perspective that the river is alive, perhaps best captured in the Whanganui Māori saying, Ko au te awa, ko te awa ko au, which means 'I am the river, and the river is me.'[7]

A similar campaign is underway around Lake Mary Jane in central Florida, which is suing to protect itself from harmful development. As reported by *The New Yorker*, 'Never before has an inanimate slice of nature tried to defend its rights in an American courtroom.' As a result, the case is facing fierce resistance, with the Florida Chamber of Commerce responding, 'Your local lake or river could sue you? Not on our watch.'[8]

It should be noted that in the USA, corporations already have legal personhood, so the argument would have to be made as to why it is more absurd for Lake Michigan to be a legal person than it is for General Motors to be. When I spoke to Michael Pollan, he pointed out that we don't have to completely commit to animist or panpsychist beliefs to enact policy changes that lead to better environmental protection:

> *Whether it's true or not, more animistic thinking would help us defend nature. And if psychedelics can contribute to that, they might have an important role to play.... Seeing the world in a panpsychist or animist way may be hardwired into us, and then we unlearn it in school and Western science kind of knocks it out of us. You know, children all have this belief, and most traditional cultures have this belief.*[9]

Pollan is making a utilitarian argument: Even if we disagree with the underlying science or philosophy of panpsychism, we can acknowledge that it has benefits as a worldview.

Decriminalization and Liberty

A shift to valuing consciousness above things could also impact drug laws themselves. In most parts of the world, psychedelics are still illegal. There are a number of movements around the world looking to decriminalize or reschedule them. One is Decriminalize Nature in the USA; as I mentioned earlier in the book, this movement drove the legalization of psychedelics in Oakland, California. In the UK, the Conservative Drug Policy Reform Group (CDPRG) is running a campaign around psilocybin access rights to reschedule psilocybin in the UK. It is led by MP Crispin Blunt, who used to be the Prisons Minister.

When I spoke to him, he shared his insight during that time was that if you can't keep drugs out of a high-security prison, it's absurd to think you can keep them out of society at large.[10] Most people now agree that the war on drugs has been a catastrophic failure. Rooted in racism and bigotry, it disproportionately affects minority populations, and destroys whole communities.

I am supportive not just of psychedelic decriminalization movements, but the decriminalization and regulation of all drugs. It is a human-rights violation to imprison people for changing their consciousness. If we orient around the quality of consciousness as our key value, then imprisoning someone for choosing to change their consciousness is a horrific crime. It is not just a physical imprisonment, but an imprisonment of the self – a limitation on who we can become, how we can think, and how we experience the world.

At the same time, with freedom comes responsibility. The entrance of LSD into culture in the 1960s brought profound changes in social attitudes for some, but also chaos, psychological damage, abuse, and cults. The complex figure responsible for some of this, Timothy Leary, argued in his book *The Politics of Ecstasy* that psychedelics could be regulated by giving people a license, as we do with driving.[11] This idea has been revived by Timmy Davis, who leads the CDPRG's campaign to reschedule psilocybin in the UK. He suggests that a licensing system could allow different practitioners to offer psilocybin, for example, and put the power in the hands of the individual to decide if they want to experience a Christian mushroom ceremony, a Shipibo-Conibo ayahuasca ceremony, or simply go into the woods to commune with nature.[12] Of course, there are still a number of questions around regulation – in particular, which authority is in charge of it, whether it's seen as a health issue or civil issue, and more.

Regardless of which models we end up adopting, more public education around the psychedelic experience is essential. People need to know how to prepare, how to trip, and how to integrate. Depending on their mental-health history, they need clear routes to access professionals, and clear and safe ways to access psychedelics for spiritual growth, recreation, and personal development.

Unlocking this freedom is like splitting an atom – it releases tremendous energy into the culture. In the 1960s, we weren't ready. We still don't have frameworks and institutions that can help us hold that kind of transformative and chaotic power. Medicalized clinics will only ever be one part of the ecosystem, and I don't believe these are suited to help people make sense of spiritual transformation. As well as clinics, we'll also need places of worship.

As I've shared, finding ways to come together and experience communitas, a collective sense of being part of something larger, is essential to our social cohesion and well-being. But there is another reason the idea of spiritual practice is so important. I would see, as my final dosing drew to a close, that we need new ways to connect, as well as the humility to remember that reality is stranger than we can guess. The psychedelic experience is inherently mysterious, and a spiritual practice can help us to navigate this mystery in meaningful ways.

First Contact

With a sense that I only had seconds left until the dosing ended, I asked the Teaching Presence one final question: 'How can we use technology like this to change our trajectory?'

In response, the Teaching Presence gave me the hyperdimensional equivalent of a shrug.

'That's up to all of you to decide. Nobody could or would ever give you the answer to that.'

'Should we start using this dosing technology to travel here more often, in order to gain insight from other intelligences?'

I received a cosmic smile in response. 'You can't even figure out how to get along with one another. What makes you think you'd be able to relate effectively with other intelligences?'

It had a point. Reflecting on this, I had the thought that the Internet is in some way our simplistic testing ground for traveling to a wider ecosystem of consciousness. But we're so far from being able to show up with virtue and integrity even in our own small version – our nursery, our playground. Thinking in these terms, I had the sudden feeling that the whole dosing experience was very sci-fi.

'Everything you're experiencing here is your overlay,' the Teaching Presence explained, 'Your mind's way of making sense of this information. You read a lot of sci-fi.'

For a brief moment, it pulled back a veil I didn't know was there and I was enveloped by something so big, so weird, so vast and mysterious, that I couldn't grasp it. A screaming abyss of infinite potential opened up all around and through me. As quickly as it appeared, the veil was back – and with it, my own cognitive and emotional framework.

The Teaching Presence had shown me, for a split second, what was behind the frame I'm using. My frame is necessary for me to make sense of the world around me, but it doesn't touch the surface of what is really going on. I was being presented with something it's painfully easy to forget as a human being: We don't know as much as we think we do. And we must find ways to consciously remember this. It is a kind of remembrance that we have traditionally found through some form of spiritual or religious practice.

A Renaissance in Religion

In 2009, I wrote an article for *The Guardian* arguing that psychedelics should be legalized for spiritual practice.[13] When the article came out, I was at a friend's house and so excited that I opened it on a laptop to watch the comments section. I witnessed in mounting horror as the most liberal readers in the

UK piled on and laid into me for being a naive hippie. As one commenter put it, '"Ooooh, I found spiritual enlightenment with a few pills/drinks/shrooms! And really, it's genuine! I really grew as a human, maaaaan!" Please.'

Secular cultures don't value spiritual experience, and in many ways, psychotherapy has come to play the role of religious intermediary in the West. It is no surprise that psychedelic medicalization has received far more attention than ceremonial practice. In a series of conversations I had with therapist and psychedelic activist Kat Conour, she pointed out that it's now inevitable that we'll see capitalist, extractive versions of a psychedelic future. That may be unavoidable in our existing system. However, we'll also see deeply held ceremonial processes. People will still use psychedelics in a ceremonial or religious setting, whether they are legal or not.[14]

Conour and others have argued that what we need to ensure is a healthy psychedelic ecosystem – one in which no single modality (for example, medical pharmacology) dominates who gets access and what kinds of experiences we can have. I have summarized this as a need for 'clinics and churches,' or the idea that the distribution of power over access needs to be balanced.[15]

Who decides who's sick? Who gets the power of determining what sickness is, to begin with? A shaman or a rabbi might diagnose a spiritual cause to someone's depression, whereas a psychiatrist might diagnose it as a physical ailment. As the psychedelic renaissance matures, we can make space for all of these interpretations, and put the choice back with the individual wherever possible.

It may be easier and safer to try and bring psychedelics into some existing religious frameworks than to create new religions from scratch. New religious movements can be exciting and empowering, but to date, few have scaled successfully. Some have caused a lot of damage on the way. As yet, there are relatively few psychedelic religions, but with so many underground ceremonial circles and spiritual communities rife with emotional and sexual abuse, the idea of new psychedelic religions should at least be approached with caution.

Religious institutions are still the primary access to the sacred for billions of people around the world. However, religious observance is decreasing globally. The argument could be made that many religions need to evolve to stay relevant to our lives today, and that they need reinvigorating.

One way to do this is to focus on creating experiences. Evangelical churches, for example, are growing in popularity.[16] They involve an experience of the divine, instead of the usual behavioral and social function of religions that people observe but don't really feel transforms who they are on a deep level. Carl Jung called this a 'creed' rather than a religion that gives us a religious experience. This is echoed by Albert Hofmann, the legendary chemist who discovered LSD, when he was asked how we can reconcile visionary experiences with religion:

> It is important to have the experience directly. Aldous Huxley taught us not to simply believe the words, but to have the experience ourselves. This is why the different forms of religion are no longer adequate. They are simply words, words, words without the direct experience of what it is the words represent...

what is of greatest importance, is that we have a personal
spiritual experience. Not words, not beliefs, but experience.[17]

As well as new religions, practitioners from existing faiths are
trying to integrate psychedelic experiences into worship. One
is Ligare, set up by Hunt Priest, a pastor who had a profound
mystical experience on a psilocybin trial for religious leaders.
Ligare is made up of 'clergy, chaplains, religious educators,
scholars, spiritual guides, philanthropists, and researchers
dedicated to bringing the direct experience of the sacred
to all who desire it through ritualized engagement with
psychedelic substances within the context of the Christian
contemplative tradition.'[18]

Another example is Divine Assembly, a psilocybin church set up
by people who left the Mormon Church. As Cassady Rosenblum
and Kim Raff report in a 2022 *Rolling Stone* article:

Although the Divine Assembly is not limited to former LDS
members, or 'post-Mormons' as they refer to themselves, the
majority of the crowd by default is, and they're aching for a
new kind of spirituality to fill the void. One couple, Yesenia
and Guillermo Ramos, tell me they left the LDS [Latter Day
Saints] Church in 2012, after it began to feel like the opposite
of what they thought it stood for. 'God is love,' Yesenia says
with conviction, but within the church, she says she felt judged
for her decision to be both a mom and a nurse, rather than a
stay-at-home mom.[19]

Yesenia's experience points to Jung's idea about creed versus
religion. The creed is restrictive, while the religious experience

is expansive. The Divine Assembly's use of psilocybin is legal due to the Religious Freedom Restoration Act, which allows worshipers to use psychedelics if the use is both safe and an integral part of their faith. The ayahuasca church União do Vegetal was granted approval on similar grounds by the US Supreme Court; so was the Native American Church's sacramental use of peyote.

These religions allow people to have direct contact with the divine. They affirm that psychedelic experiences are sacred for many reasons – and perhaps the most striking is that they allow us to die before we die.

Death and Rebirth

The mystic George Gurdjieff had a teaching that might be summed up as, 'The most important thing to remember in life is that you, and everyone you meet today, will soon be dead.'[20] As I explored earlier in the book, the process of allowing the old to die so the new can be born is essential for both individual and cultural transformation.

Psychedelics can have a profound effect on how we perceive death, and our willingness to embrace it. A 2016 study led by Roland Griffiths treating terminally ill people with psilocybin yielded astonishing results.[21] According to the group's press release:

> The Johns Hopkins group reported that psilocybin decreased clinician- and patient-rated depressed mood, anxiety and death anxiety, and increased quality of life, life meaning and optimism. Six months after the final session of treatment, about 80 percent of participants continued to show clinically

significant decreases in depressed mood and anxiety, with about 60 percent showing symptom remission into the normal range. Eighty-three percent reported increases in well-being or life satisfaction. Some 67 percent of participants reported the experience as one of the top five meaningful experiences in their lives, and about 70 percent reported the experience as one of the top five spiritually significant lifetime events.[22]

Lauren Macdonald is a medical doctor and a founding member of the Conscious Death Collective, a UK-based group of facilitators, nurses, doulas, doctors, and space holders, brought together by a mutual interest in the potential of psychedelics for reframing death and dying. Years before she became a psychedelic advocate, Macdonald was diagnosed with Stage IV cancer, and went on a healing journey that eventually led her to a legal psilocybin mushroom retreat in the Netherlands. She describes her experience:

People from my past are present but not in human form; instead, they exist as beautiful coloured swirling energies. Somehow they share the insight that dying is not final, simply the end of our physical form. I am finally at peace, surrounded by love. In the distance, I hear the quiet sound of mournful crying. I sense a presence by my side – a facilitator has come to comfort me – and I realise that I am crying. I am supported and safe as my tears start to flow. Years of pain released at last.[23]

Sometimes these experiences are profoundly transpersonal, and sometimes they can be more psychological. Kerry Pappas was a participant in the 2016 Johns Hopkins study. Diagnosed with Stage III lung cancer in 2013, Pappas experienced crippling anxiety. In an interview on *60 Minutes*, she describes her dosing: '[I was in] an ancient, prehistoric, barren land. And there's these men with pickaxes just slamming on the rocks... it felt absolutely real. I was being shown the truth of reality. Life is meaningless; we have no purpose. And then I look... and see a beautiful, shimmering bright jewel. There was sound, and it was booming. [It said,] "Right here, right now, you are alive. Right here, right now, because that's all you have." And that is my mantra to this day.'[24]

Despite the cancer having spread at the time of the documentary, Pappas had overcome her anxiety, like many on the study. 'It's amazing,' she told the presenter. 'I feel like death doesn't frighten me, and living doesn't frighten me.'

From the perspective of a conscious universe, it may be that psychedelics are healing because they reveal to us the ultimate truth about our existence: Death is not the end. As Bernardo Kastrup told me: 'The things [we accumulate] stay behind when that major change in your state of consciousness that we call death happens. The things stay behind, but what's inside goes with you. They are the only things that you carry with you because they become part of you, and you don't disappear. Where are you going to disappear? If your mind is the mind of nature, where is it going to disappear into? Where is it going to go?'[25]

What I find particularly touching and important about the power of psychedelics in palliative care is that they bring so many people to acceptance. In a culture that tells us to hold on to things, we have very little wisdom about how to let go, how to step aside. This capacity, which I explored in Chapter Two, is essential in dealing with complexity, and it appears to be essential in reconciling us to our own mortality, as well.

Eastern spiritual traditions like Buddhism teach us how to let go of control, but we often get tangled up in the West when we try to apply them to our individualist, control-focused culture. However, the West might be advised to return to its own roots. Peter Kingsley has argued that Western philosophy and all the science that sprung from an ancient Greek tradition had links to Asian shamanism and practices of death and rebirth. A core practice was incubation, a deep trance state used throughout the ancient world and in some modern Sufi practices today. It can be described as a deep death meditation.[26]

One inheritor of this tradition was the mystic-philosopher Empedocles, who preceded Plato. Empedocles believed that we live our lives deceiving ourselves, forgetting we and everything around us are eternal and divine. And because of this, he taught, we have absolutely no idea what we're doing. We get ourselves into all sorts of tangles of our own making, just as we're facing now with the Big Crisis.

In ancient Greek, there was a word for cunning and wherewithal, mêtis, that appears in both Empedocles' and Plato's writings. We could call it the ultimate ability to master how we make sense of reality, how we see what it truly is. In the tradition that lies at the heart of Western and Eastern culture, true mêtis comes only through surrendering; through recognizing that the guidance and knowledge we need come from outside of us – from a divine source.

Surrender is an essential component of the psychedelic experience. It invites us to let go, to recognize our powerlessness;

in doing so, we regain our true power. If we can learn how to surrender to complexity first, and build our laws and technologies from there, we will usher in a very different kind of world.

Conclusion:
Coming Back to Earth

As my final dose started to dim, once again I saw spread before me the vast ecosystem of consciousness – teeming with life and possibility, aching with a meaning and purpose I couldn't grasp. And it seemed clear, from that vantage point, that human beings have a choice. We can figure out how to grow up as a species, or not. Growing up is going to be painful. It's going to take work, determination, hard conversations, deep humility, fearless individualism, collective sacrifice, singing and dancing and laughter. It's going to take everything all at once.

Throughout this book, I have drawn on psychedelic research and philosophy to show how these medicines can help us in that journey; how the way they inspire us to think and perceive can be applied to making sense of the bigger picture. Throughout, I have also tried to hold the mystery of these substances – not just what they can show us, but the questions they invite us to ask about the times we live in, our strange online worlds, and the nature of reality itself.

As this book draws to a close, my hope is to leave you not just with new perspectives, but this sense of mystery. We don't really understand the psychedelic experience, and perhaps the fact that we can't is one of its most important teachings.

As I write this, I have the words of the Teaching Presence echoing through my mind from when I asked it what it was.

You but not you, we but not we.

Those words have led me to the idea that while it's up to all of us to find a way to change collectively, we aren't alone. We have each other. If we can approach psychedelics with the reverence and humility that the sacred deserves, we may also have access to profound wisdom beyond ourselves.

This may be the deepest truth psychedelic medicines can teach us: that what we need to change can't be drawn from the same soil that's making us sick. It can't be found inside the narrow frames that are constricting our thought. It is found both within and beyond us – down in the depths of our own truths, and somewhere far beyond our greatest imaginings.

As I have explored throughout the book, any collective healing requires new ways for us to come together, new ways to talk about difficult topics, new ways to disagree, and new ways to coordinate. As religious scholar and Good Friday experiment participant Huston Smith said, 'If you think you're drawing closer to God, and not in fact drawing closer to your neighbour, you're just fooling yourself.'[1]

This brings us to one last lesson I'd like to draw from the psychedelic experience. It's a lesson in how to care for one another. For me, what is most profound about the psychedelic experience is that it is an encounter with something beyond us that cares deeply about our healing; something mysterious and vast that loves us. For so many, it is an experience defined by an all-encompassing love and wisdom. It wants us to get better, and asks nothing of us in return. This kind of love can be an inspiration, if we let it. It is the creative force that comes right from the heart of the human spirit and the universe itself. Perhaps the most radical revolution we can enact is to embody some part of this love in our day-to-day lives; in how we show up with one another; in how we treat ourselves; in how we care for the sacred land we all share.

This creative love is what we will need to traverse the Big Crisis together. It is the new soil from which we can draw the grit, the collaboration, the humor and inspiration that we need to process our collective traumas and shadows. Psychedelics open the doors of perception, but it's up to us to step through. As these medicines go mainstream around the world, we have an opportunity to decide together how to do that with the courage, compassion, and virtue that we all deserve. If we can, perhaps we will create a new reality beyond our wildest imaginations.

Endnotes

Introduction

1. Grof S, Bennett HZ. The holotropic mind: The three levels of human consciousness and how they shape our lives. Reprint edition. San Francisco: HarperOne; 2009.

2. Griffiths RR, Richards WA, McCann U, Jesse R. Psilocybin can occasion mystical-type experiences having substantial and sustained personal meaning and spiritual significance. Psychopharmacology (Berl). 2006 Aug 18; 187(3):268–92.

3. CDPRG. Public throws support behind leading scientists' calls to change UK regulations on psychedelic drug research. [Internet]. Conservative Drug Policy Reform Group 2021 Jun 4 [cited 2022 Nov 18]. Available from: https://www.cdprg.co.uk/blog/psilonautica-drugscience-yougov-psilocybin-results

4. Khan F. We are engaged in something much more complex than a debate about the evidence [Internet]. The British Psychological Society 2022 Aug 16 [cited 2022 Nov 18]. Available from: https://www.bps.org.uk/psychologist/we-are-engaged-something-much-more-complex-debate-about-evidence

5. Ltd IAP. Psychedelic therapeutics market worth $ 8.31 billion by 2028 – exclusive report by InsightAce Analytic. [Internet]. [cited 2022 Nov 1]. Available from: https://www.prnewswire.co.uk/news-releases/psychedelic-therapeutics-market-worth-8-31-billion-by-2028-exclusive-report-by-insightace-analytic-816360435.html

Chapter One

1. Beiner A. The psychedelic trojan horse [Internet]. Rebel Wisdom 2021 [cited 2022 Nov 1]. Available from: https://medium.com/rebel-wisdom/the-psychedelic-trojan-horse-14c9704efd4

2. Davis E. Interview for The bigger picture. [Personal interview, 2022 Apr] Zoom; 2022 (unpublished).

3. Baum D. Smoke and mirrors: the war on drugs and the politics of failure. Reprint edition. Boston: Back Bay Books; 1997. p. 40.

4. Griffiths RR, Richards WA, McCann U, Jesse R. Psilocybin can occasion mystical-type experiences having substantial and sustained personal meaning and spiritual significance. Psychopharmacology (Berl). 2006 Aug 18; 187(3):268–92.

5. Doblin R. Pahnke's 'Good Friday experiment': a long-term follow-up and methodological critique. Journal of Transpersonal Psychology. 1991; 23(1):1–28.

6. Cicero MT. The political works of Cicero: Treatise on the republic and Treatise on the laws. 1st edition. Sr PAB, editor. Dallas: Veritatis Splendor Publications; 2014, p. 347.

7. Muraresku B. The immortality key: the secret history of the religion with no name. New York: St. Martin's Press; 2020.

8. Doyle R. Darwin's pharmacy: sex, plants, and the evolution of the noosphere. Seattle: University of Washington Press; 2011. p. 20.

9. Ep. 1 – Awakening from the meaning crisis – introduction [Internet]. 2019 [cited 2022 Nov 1]. Available from: https://www.youtube.com/watch?v=54l8_ewcOlY

10. Lewis M. Brain change in addiction as learning, not disease. The New England Journal of Medicine. 2018 Oct 18; 379(16):1551–60.

11. Ep. 13 – Awakening from the meaning crisis – Buddhism and parasitic processing [Internet]. 2019 [cited 2022 Nov 1]. Available from: https://www.youtube.com/watch?v=vGB8k7jk1AQ

12. Eisner B. Set, setting, and matrix. Journal of Psychoactive Drugs. 1997 Jun 29; 29(2):213–16.

13. Olson DE. Biochemical mechanisms underlying psychedelic-induced neuroplasticity. Biochemistry. 2022 Jan 21; 61(3):127–36.

14. How do psychedelics work? Robin Carhart-Harris [Internet]. 2021 [cited 2022 Nov 1]. Available from: https://www.youtube.com/watch?v=TufuI17wQ10

15. Beiner A. The psychedelic trojan horse [Internet]. Rebel Wisdom 2021 [cited 2022 Nov 1]. Available from: https://medium.com/rebel-wisdom/the-psychedelic-trojan-horse-14c9704efd4

16. Luke D. DMT dialogues: encounters with the spirit molecule. 1st edition. Rochester, VT: Park Street Press; 2018. p. 46–48.

17. Pochettino ML, Cortella AR, Ruiz M. Hallucinogenic snuff from Northwestern Argentina: microscopical identification of anadenanthera colubrina var. cebil (fabaceae) in powdered archaeological material. Econ Bot. 1999 Apr 1; 53(2):127–32.

18. Hancock G. Supernatural: meetings with the ancient teachers of mankind. New edition. Cornerstone Digital; 2010.

19. Nichols DE. N,N-dimethyltryptamine and the pineal gland: separating fact from myth. Journal of Psychopharmacology. 2018 Jan 1; 32(1):32–36.

20. Barker SA. N, N-Dimethyltryptamine (DMT), an endogenous hallucinogen: past, present, and future research to determine its role and function. Frontiers in Neuroscience. 2018 August 6.

21. Strassman R. DMT: the spirit molecule: a doctor's revolutionary research into the biology of near-death and mystical experiences. Rochester, VT: Park Street Press; 2000. p. 188.

22. Ibid, p. 190.

23. Ibid, p. 193.

24. Ibid, p. 195.

25. Davis AK, Clifton JM, Weaver EG, Hurwitz ES, Johnson MW, Griffiths RR. Survey of entity encounter experiences occasioned by inhaled N,N-dimethyltryptamine: phenomenology, interpretation, and enduring effects. Journal of Psychopharmacology. 2020 Sep 3; 34(9):1008–20.

26. Lawrence DW, Carhart-Harris R, Griffiths R, Timmermann C. Phenomenology and content of the inhaled N, N-dimethyltryptamine (N, N-DMT) experience. Scientific Reports. 2022 May 24; 12(1):8562.

27. Four Forces Framework [Internet]. The Four Forces [cited 2022 Nov 2]. Available from: https://www.thefourforces.com/four-forces-framework/

28. Almaas AH. The point of existence: transformations of narcissism in self-realization. Re-issue edition. Boulder, CO: Shambhala; 1998. p. 29.

29. Strassman R. DMT and the soul of prophecy: a new science of spiritual revelation in the Hebrew Bible. 1st edition. Rochester, VT: Park Street Press; 2014.

30. Bogenschutz MP, Ross S, Bhatt S, Baron T, Forcehimes AA, Laska E, et al. Percentage of heavy drinking days following psilocybin-assisted psychotherapy vs placebo in the treatment of adult patients with alcohol use disorder: a randomized clinical trial. JAMA Psychiatry. 2022 Oct 1; 79(10):953–62.

31. Tversky B. Mind in motion: how action shapes thought. New York: Basic Civitas Books; 2019.

32. Cardinali L, Zanini A, Yanofsky R, Roy AC, de Vignemont F, Culham JC, et al. The toolish hand illusion: embodiment of a tool based on similarity with the hand. Scientific Reports. 2021 Jan 21; 11(1):2024.

33. Spivey, M. When objects become extensions of you [Internet]. The MIT Press Reader 2020 [cited 2022 Nov 1]. Available from: https://thereader.mitpress.mit.edu/when-objects-become-extensions-of-you/

34. Embodiment & flow week 1: live recording [Internet]. 2022 [cited 2022 Nov 2]. Available from: https://www.youtube.com/watch?v=vJBaX9piJHM

35. Beiner A. The psychedelic trojan horse [Internet]. Rebel Wisdom 2021 [cited 2022 Nov 1]. Available from: https://medium.com/rebel-wisdom/the-psychedelic-trojan-horse-14c9704efd4

36. Terence McKenna – the purpose of psychedelics [Internet]. 2014 [cited 2022 Nov 1]. Available from: https://www.youtube.com/watch?v=rXQfE3W4wVY

Chapter Two

1. Rowson J, editor. Dispatches from a time between worlds: crisis and emergence in metamodernity: 1. London: Perspectiva; 2021. p. 17.

2. Pinker S. Enlightenment now: the case for reason, science, humanism, and progress. 1st edition. New York: Penguin Random House; 2018.

3. 1. Ep. 1 – Awakening from the meaning crisis – Introduction [Internet]. 2019 [cited 2022 Nov 1]. Available from: https://www.youtube.com/watch?v=54l8_ewcOlY

4. Voter turnout trends around the world [Internet]. International Institute for Democracy and Electoral Assistance 2016 Dec 31 [cited 2022 Nov 2]. Available from: https://www.idea.int/es/publications/catalogue/voter-turnout-trends-around-world

5. Memetic tribes of culture war 2.0 [Internet]. The Stoa [cited 2022 Nov 2]. Available from: https://medium.com/s/world-wide-wtf/memetic-tribes-and-culture-war-2-0-14705c43f6bb

6. Waida M. Problems of Central Asian and Siberian shamanism. Numen. 1983 Dec; 30(2):215–39.

7. Kopenawa D, Albert B, Elliott N, Dundy A. The falling sky: words of a Yanomami shaman. Illustrated edition. Cambridge, MA: Harvard University Press; 2013, p. 93.

8. Wilber K. Trump and a post-truth world. Boulder, CO: Shambhala; 2017.

9. Antonovsky A. Unraveling the mystery of health: how people manage stress and stay well. 1st edition. San Francisco: Jossey-Bass; 1987.

10. Myth, wisdom & pandemic, Stephen Jenkinson, Zak Stein & Charlotte Du Cann [Internet]. 2020 [cited 2022 Nov 2]. Available from: https://www.youtube.com/watch?v=m_c2-Bs1BqM

11. Psychedelics in a changing world: medicalisation, reciprocity and planetary healing: specialist panel recorded at Medicine Festival [Internet]. Chasing Consciousness 2021 Aug 28 [cited 2022 Nov 2]. Available from: https://www.chasingconsciousness.net/psychedelics-medicalisation-reciprocity-panel-medicinefestival2021

12. Huxley A, Ballard JG. The doors of perception: and heaven and hell. 1st edition. London: Vintage Classics; 2004.

13. Rowson J, editor. Dispatches from a time between worlds: Crisis and emergence in metamodernity: 1. London: Perspectiva; 2021. p. 31.

14. Björkman T. Interview for The bigger picture. [Personal interview, 2022 May] Zoom; 2022 (unpublished).

15. Holland J. Hidden order: how adaptation builds complexity. New York: Basic Books; 1996.

16. Henrich J. The secret of our success: how culture is driving human evolution, domesticating our species, and making us smarter. Princeton, NJ: Princeton University Press; 2015.

17. Carhart-Harris R, Leech R, Hellyer P, Shanahan M, Feilding A, Tagliazucchi E, et al. The entropic brain: a theory of conscious states informed by neuroimaging research with psychedelic drugs. Frontiers in Human Neuroscience [Internet]. 2014 [cited 2022 Nov 2]; 8. Available from: https://www.frontiersin.org/articles/10.3389/fnhum.2014.00020

18. Korzybski A, Pula RP. Science and sanity: an introduction to non-Aristotelian systems and general semantics. 5th edition. New York: Institute of General Semantics; 1995.

19. Morton T. Hyperobjects: philosophy and ecology after the end of the world. Minneapolis: University of Minnesota Press; 2013.

20. McGilchrist I. The master and his emissary: the divided brain and the making of the Western world. 2nd edition. New Haven, CT: Yale University Press; 2019.

21. McGilchrist I. The matter with things: our brains, our delusions, and the unmaking of the world. London: Perspectiva; 2021.

22. McGilchrist I. Interview for The bigger picture. [Personal interview, 2022 Jul] Zoom; 2022 (unpublished).

23. Serafetinides EA. The EEG effects of LSD-25 in epileptic patients before and after temporal lobectomy. Psychopharmacologia. 1965 May 21; 7(6):453–60.

24. Jaynes J. The origin of consciousness in the breakdown of the bicameral mind. Boston New York: Houghton Mifflin; 2000.

25. Strassman R. DMT: the spirit molecule: a doctor's revolutionary research into the biology of near-death and mystical experiences. Rochester, VT: Park Street Press; 2000. p. 275.

26. The war on sensemaking, Daniel Schmachtenberger [Internet]. 2019 [cited 2022 Nov 2]. Available from: https://www.youtube.com/watch?v=7LqaotiGWjQ

27. The war on pineapple: understanding foreign interference in 5 steps [Internet]. Cybersecurity and Infrastructure Security Agency 2019 Jul [cited 2022 Nov 2]. Available from: https://www.cisa.gov/sites/default/files/publications/19_1008_cisa_the-war-on-pineapple-understanding-foreign-interference-in-5-steps.pdf

28. Miller C. The death of the gods: the new global power grab. 2nd edition. Cornerstone Digital; 2018.

29. Can truth survive big tech? Tristan Harris [Internet]. 2020 [cited 2022 Nov 2]. Available from: https://www.youtube.com/watch?v=wHQQFOv7QgQ

30. Fuller D. The uncanny valley: medicines, censorship & the problem of truth [Internet]. Rebel Wisdom 2021 [cited 2022 Nov 2]. Available from: https://medium.com/rebel-wisdom/the-uncanny-valley-medicines-censorship-the-problem-of-truth-f4cac2ffebd9

31. Rao V. The Internet of beefs [Internet]. Noema/Berggruen Institute 2020 Sep 22 [cited 2022 Nov 2]; Available from: https://www.noemamag.com/the-internet-of-beefs

32. Haidt J. Why the past 10 years of American life have been uniquely stupid [Internet]. The Atlantic. 2022 Apr 11 [cited 2022 Nov 2]. Available from: https://www.theatlantic.com/magazine/archive/2022/05/social-media-democracy-trust-babel/629369/

33. UN/DESA policy brief #108: Trust in public institutions: trends and implications for economic security [Internet]. Department of Economic and Social Affairs [cited 2022 Nov 2]. Available from: https://www.un.org/development/desa/dpad/publication/un-desa-policy-brief-108-trust-in-public-institutions-trends-and-implications-for-economic-security/

34. Henrich J. The weirdest people in the world: how the West became psychologically peculiar and particularly prosperous. 1st edition. New York: Penguin Random House; 2020. p. 21.

35. The weirdest people in the world with Joseph Henrich [Internet]. 2021 [cited 2022 Nov 2]. Available from: https://www.youtube.com/watch?v=ubKZ0IbX7Ic

36. Eisenstein C. Climate: a new story. Berkeley, CA: North Atlantic Books; 2018.

37. Yunkaporta T. Sand talk: how indigenous thinking can save the world. San Francisco: HarperOne; 2019.

38. A startling new book from Australia – 'Sand talk: how indigenous thinking can save the world' – gives lessons in complexity [Internet]. The Alternative 2020 18 Jul [cited 2022 Nov 2]. Available from: https://www.thealternative.org.uk/dailyalternative/2020/7/19/sand-talk

39. Pollan M. This is your mind on plants: opium – caffeine – mescaline. New York: Penguin Random House; 2021.

40. Richards W. Sacred knowledge: psychedelics and religious experiences. New York: Columbia University Press; 2015.

41. The forward escape – Terence McKenna [Internet]. 2016 [cited 2022 Nov 2]. Available from: https://www.youtube.com/watch?v=Hr3f6gIz0-k

42. Ep. 37 – Awakening from the meaning Crisis – Reverse engineering enlightenment: part 2 [Internet]. 2019 [cited 2022 Dec 14]. Available from: https://www.youtube.com/watch?v=2kQooMZzR7w

43. Hilbert M. How much information is there in the "information society"? Significance. 2012 August 9; 9(4):8–12.

44. Tolle E. The power of now: a guide to spiritual enlightenment. Novato, CA: New World Library; 2010.

45. Porges SW. Polyvagal theory: a science of safety. Frontiers in Integrative Neuroscience. 2022 May 10; 16.

46. Bonanno G. The end of trauma: how the new science of resilience is changing how we think about PTSD. New York: Basic Books; 2021.

47. Murphy-Beiner A, Soar K. Ayahuasca's 'afterglow': improved mindfulness and cognitive flexibility in ayahuasca drinkers. Psychopharmacology. 2020 Apr 1; 237(4):1161–69.

48. Wießner I, Falchi M, Maia LO, Daldegan-Bueno D, Palhano-Fontes F, Mason NL, et al. LSD and creativity: Increased novelty and symbolic thinking, decreased utility and convergent thinking. Journal of Psychopharmacology. 2022 Mar 1; 36(3):348–59.

49. Ep. 13 – Awakening from the meaning crisis – Buddhism and parasitic processing [Internet]. 2019 [cited 2022 Dec 14]. Available from: https://www.youtube.com/watch?v=vGB8k7jk1AQ

Chapter Three

1. Newson M, Khurana R, Cazorla F, van Mulukom V. 'I get high with a little help from my friends' – how raves can invoke identity fusion and lasting co-operation via transformative experiences. Frontiers in Psychology [Internet]. 2021 Sep 24 [cited 2022 Nov 9]; 12. Available from: https://www.frontiersin.org/articles/10.3389/fpsyg.2021.719596

2. Forstmann M, Yudkin DA, Prosser AMB, Heller SM, Crockett MJ. Transformative experience and social connectedness mediate the mood-enhancing effects of psychedelic use in naturalistic settings. Proceedings of the National Academy of Sciences. 2020 Feb 4; 117(5):2338–46.

3. Walsh R. The world of shamanism: new views of an ancient tradition. 1st Edition. Woodbury, MN: Llewellyn; 2007. p. 209–11.

4. Williams M, Negrin D, Eriacho B, Buchanan N, Davis E, de Leon C. Psychedelic Justice: Toward a Diverse and Equitable Psychedelic Culture. Labate BC, Cavnar C, editors. Synergetic Press; 2021, pp. 84.

5. Walsh R. The world of shamanism: new views of an ancient tradition. 1st Edition. Woodbury, MN: Llewellyn; 2007. p. 209.

6. Gonzalez D, Cantillo J, Perez I, Carvalho M, Aronovich A, Farre M, et al. The Shipibo ceremonial use of ayahuasca to promote well-being: an observational study. Frontiers in Pharmacology. 2021 May 5; 12.

7. Kettner H, Rosas FE, Timmermann C, Kärtner L, Carhart-Harris RL, Roseman L. Psychedelic communitas: intersubjective experience during psychedelic group sessions predicts enduring

changes in psychological wellbeing and social connectedness. Frontiers in Pharmacology. 2021 Mar 25; 12.

8. Kotler S, Wheal J. Stealing fire: how Silicon Valley, the Navy SEALs, and maverick scientists are revolutionizing the way we live and work. Reprint edition. New York: Dey Street Books; 2017.

9. Putnam R. Bowling alone: the collapse and revival of American community. New edition. London: Simon & Schuster; 2001.

10. Beiner A. The intimacy crisis: sex and social media with Katherine Dee [Internet]. Rebel Wisdom 2022 [cited 2022 Nov 9]. Available from: https://rebelwisdom.substack.com/p/the-intimacy-crisis-sex-and-social

11. John Della Volpe [@dellavolpe]. NEW POLLING I conducted and shared w/ @Morning_Joe team on social media, #Facebook, #Instagram. 1) Nearly 2/3 of Americans who use platforms believe life was better without them. 2) 42% of #GenZ addicted, can't stop if they tried. 1/5 https://t.co/iHwM30puj5 [Internet]. Twitter 2021 Oct 12 [cited 2022 Nov 9]. Available from: https://twitter.com/dellavolpe/status/1448022068193333248/photo/1

12. Beiner A. The intimacy crisis: sex and social media with Katherine Dee [Internet]. Rebel Wisdom 2022 [cited 2022 Nov 9]. Available from: https://rebelwisdom.substack.com/p/the-intimacy-crisis-sex-and-social

13. Turkle S. Alone together: why we expect more from technology and less from each other. New York: Basic Books; 2017.

14. Harrington M. Welcome to fully automated luxury gnosticism [Internet]. UnHerd 2021 Sep 23 [cited 2022 Nov 9]. Available from: https://unherd.com/2021/09/welcome-to-fully-automated-luxury-gnosticism/

15. Turkle S. Alone together: why we expect more from technology and less from each other. New York: Basic Books; 2017.

16. Ibid, p. 28–35.

17. Nguyen CT. Games: agency as art. Oxford: Oxford University Press; 2020.

18. Nguyen CT. How Twitter gamifies communication. In: Lackey J, editor. Applied Epistemology [Internet]. Oxford: Oxford University Press; 2021. p. 410–36.

19. Nguyen CT. Games: agency as art. Oxford: Oxford University Press; 2020.

20. Schmachtenberger D. Technology is not values neutral: ending the reign of nihilistic design [Internet]. The Consilience Project 2022 Jun 26 [cited 2022 Nov 9]. Available from: https://consilienceproject.org/technology-is-not-values-neutral/

21. Harrison K. Psychedelic mysteries of the feminine: creativity, ecstasy, and healing. Papaspyrou M, Baldini C, Luke D, editors. Rochester, VT: Park Street Press; 2019. p. 145.

22. Lemoine B. May be fired soon for doing AI ethics work [Internet]. Medium 2022 Jun 6 [cited 2022 Nov 9]. Available from: https://cajundiscordian.medium.com/may-be-fired-soon-for-doing-ai-ethics-work-802d8c474e66

23. Opinion | Did Google create a sentient program? [Internet]. MSNBC.com 2022 Jun 17 [cited 2022 Nov 9]. Available from: https://www.msnbc.com/opinion/msnbc-opinion/google-s-ai-impressive-it-s-not-sentient-here-s-n1296406

24. Luke D. DMT dialogues: encounters with the spirit molecule. 1st edition. Rochester, VT: Park Street Press; 2018. p. 71.

25. Ibid, p. 73.

26. Memes: virality and the occult w/ Chris Gabriel (MemeAnalysis) [Internet]. 2020 [cited 2022 Dec 14]. Available from: https://www.youtube.com/watch?v=V4oc9tqK_eM

27. Jung CG. The archetypes and the collective consciousness. 2nd edition, reprint. London: Routledge; 2007. p. 48.

28. Grof S. Mythic imagination and modern society: the re-enchantment of the world. In: 16th International Transpersonal Association Conference [Internet]; 2004 Jun 12–18; Palm Springs, CA. Available from: https://www.stangrof.com/images/joomgallery/ArticlesPDF/ITA_Palm_Springs-archetypes_rev01-2014.pdf

29. Harner M. The way of the shaman. 10th anniversary edition. San Francisco: HarperOne; 2011.

30. Luke D. DMT dialogues: encounters with the spirit molecule. 1st edition. Rochester, VT: Park Street Press; 2018. p. 131.

31. Hancock G. Supernatural: meetings with the ancient teachers of mankind. New edition. Cornerstone Digital; 2010.

32. Luke D. Interview for The bigger picture. [Personal interview, 2022 Apr] In person; 2022 (unpublished).

33. Read T. Interview for The bigger picture. [Personal interview, 2022 Jul] In person; 2022 (unpublished).

34. Read T. Walking shadows: archetype and psyche in crisis and growth. New edition. London: Aeon; 2019. p. 47–48.

35. Lyons NS. The Reality War [Internet]. The Upheaval 2021 Oct 29 [cited 2022 Nov 9]. Available from: https://theupheaval.substack. com/p/the-reality-war

36. Inglehart RF. Giving up on God: the global decline of religion. Foreign Affairs. 2020 Aug 11; 5:112–18.

37. Burton TI. Strange rites: new religions for a godless world. New York: Public Affairs; 2020. p. 10.

38. Jung CG. The Earth has a soul: C.G. Jung's writings on nature, technology and modern life. 1st edition. Sabini M, editor. Berkeley, CA: North Atlantic Books; 2002.

39. Kingsnorth P. The cross and the machine [Internet]. First Things 2021 Jun [cited 2022 Nov 9]. Available from: https://www. firstthings.com/article/2021/06/the-cross-and-the-machine

40. Bourgeault C. Introducing the imaginal [Internet] 2018 Nov 13 [cited 2022 Nov 9]. Available from: https://cynthiabourgeault. org/2018/11/13/introducing-the-imaginal/

41. Sullivan A. The strange rebirth of Imperial Russia [Internet]. The Weekly Dish 2022 Mar 25 [cited 2022 Nov 9]. Available from: https://andrewsullivan.substack.com/p/the-strange-rebirth -of-imperial-russia

42. Will be wild. [Podcast]. Wondery 2022 25 April [cited 9 November]. Available from: https://wondery.com/shows/will-be-wild/

43. Ward C, Voas D. The emergence of conspirituality. Journal of Contemporary Religion. 2011 Jan 7; 26(1):103–21.

44. Evans J. Interview for The bigger picture. [Personal interview, 2022 Apr] Zoom; 2022 (unpublished).

45. Wheal J. Recapture the rapture: rethinking God, sex, and death in a world that's lost its mind. New York: Harper Wave; 2021.

46. Conspiracy & pandemic, with John Vervaeke, Carl Miller & Jules Evans [Internet]. 2020 [cited 2022 Nov 9]. Available from: https://www.youtube.com/watch?v=1P8pLXYT44U

47. QAnon candidates are on the ballot in 26 states [Internet]. Grid News 2022 Apr 12 [cited 2022 Nov 9]. Available from: https://www.grid.news/story/misinformation/2022/04/12/qanon-candidates-are-on-the-ballot-in-26-states/

48. Luke D. DMT dialogues: encounters with the spirit molecule. 1st edition. Rochester, VT: Park Street Press; 2018. p. 79.

49. Ibid. p. 80.

50. Thompson HS. Fear and loathing on the campaign trail '72. New York: Harper Perennial; 2014. p. 538.

51. The Jazz Leadership Project [Internet]. [cited 2022 Dec 14]. Available from: https://www.jazzleadershipproject.com/

52. Bartlett R. Interview for The bigger picture. [Personal interview, 2022 Apr] Zoom; 2022 (unpublished).

53. Thomas G. Antagonistic cooperation: advancing through challenge [Internet]. Tune Into Leadership 2022 Jan 6 [cited 2022 Dec 14]. Available from: https://www.tuneintoleadership.com/blog/antagonistic-cooperation-advancing-through-challenge

54. Hamilton DM, Wilson GM, Loh KM. Compassionate conversations: how to speak and listen from the heart. Boulder, CO: Shambhala; 2020.

55. Ness S. Course workbook. [Online course: The art of difficult conversations 2022 Apr] Zoom; 2022 (unpublished).

56. Porges SW. Polyvagal theory: a science of safety. Frontiers in Integrative Neuroscience. 2022 May 10; 16.

57. Vervaeke J. Interview for The bigger picture. [Personal interview, 2022 Apr] Zoom; 2022 (unpublished).

Chapter 4

1. Bisbee CC, Bisbee P, Dyck E, Farrell P, editors. Psychedelic prophets: the letters of Aldous Huxley and Humphry Osmond. 3rd edition. Montreal/Kingston, ON: McGill-Queen's University Press; 2018. p. 266–67.

2. The whole spectrum of shadows self-paced course level 01 Integral framework [Internet]. [cited 2022 Nov 26]. Available from: https://integralzen.org/event/the-whole-spectrum-of-shadows-self-paced-course-level-1

3. Shulgin A. Psychedelic psychotherapy and the shadow. In: Mind States Conference [Internet]; 2002 Oct 1–6; Negril, Jamaica. Available from: http://www.matrixmasters.net/podcasts/TRANSCRIPTS/AnnShulgin-TheShadow1.html

4. Morgan R. Interview for The bigger picture. [Personal interview, 2022 May] In Person; 2022 (unpublished).

5. Shulgin A. Psychedelic psychotherapy and the shadow. In: Mind States Conference [Internet]; 2002 Oct 1–6; Negril, Jamaica. Available from: http://www.matrixmasters.net/podcasts/TRANSCRIPTS/AnnShulgin-TheShadow1.html

6. Solzhenitsyn A. The gulag archipelago [volume 1]: an experiment in literary investigation. Reissue edition. New York: Harper Perennial; 2020. p. 168.

7. Shulgin A. Psychedelic psychotherapy and the shadow. In: Mind States Conference [Internet]; 2002 Oct 1–6; Negril, Jamaica. Available from: http://www.matrixmasters.net/podcasts/TRANSCRIPTS/AnnShulgin-TheShadow1.html

8. Roseman L, Haijen E, Idialu-Ikato K, Kaelen M, Watts R, Carhart-Harris R. Emotional breakthrough and psychedelics: validation of the emotional breakthrough inventory. Journal of Psychopharmacology. 2019 Sep 1; 33(9):1076–87.

9. Jung CG. Civilization in transition. 1st edition. Fordham M, Read SH, Adler G, editors. London: Routledge; 1964. p. 164.

10. Krause I. What is global social witnessing? [Internet]. Thomas Hübl. 2019 [cited 2022 Nov 14]. Available from: https://thomashuebl.com/what-is-global-social-witnessing/

11. Roseman L, Ron Y, Saca A, Ginsberg N, Luan L, Karkabi N, et al. Relational processes in ayahuasca groups of Palestinians and Israelis. Frontiers in Pharmacology. 2021 19 May; 12.

12. Roseman L. Interview for The bigger picture. [Personal interview, 2022 Aug] In Person; 2022 (unpublished).

13. Roseman L, Karkabi N. On revelations and revolutions: drinking ayahuasca among Palestinians under Israeli occupation. Frontiers in Psychology. 2021 August 27; 12.

14. Roseman L. Additional notes for revelations and revolutions: drinking ayahuasca among Palestinians under Israeli occupation. Unpublished.

15. Didion J. Slouching towards Bethlehem: essays. New York: Open Road Media; 2017. p. 100.

16. Dohrn Z. Mother country radicals. [Podcast]. Crooked Media 2022 Jun 23 [cited 2022 Nov 14]. Available from: https://crooked.com/podcast/chapter-5-new-morning/

17. Ginsberg A. Collected poems 1947–1997. New York: Penguin; 2013. p. 134.

18. Meditations on Moloch [Internet]. Slate Star Codex 2014 Jul 30 [cited 2022 Nov 14]. Available from: https://slatestarcodex.com/2014/07/30/meditations-on-moloch/

19. A poker pro explains game theory: Liv Boeree [Internet]. 2021 [cited 2022 Nov 14]. Available from: https://www.youtube.com/watch?v=p-r54G30kbo

20. McCarraher E. The enchantments of Mammon: how capitalism became the religion of modernity. Cambridge, MA: Belknap Press; 2019.

21. Steinmetz-Jenkins D. Has capitalism become our religion? [Internet]. The Nation 2019 Oct 4 [cited 2022 Nov 15]. Available from: https://www.thenation.com/article/archive/capitalism-religion-eugene-mccarraher-interview/

22. Davis E. Interview for The bigger picture. [Personal interview, 2022 June] In Person; 2022 (unpublished).

23. Gupta A, Howell ST, Yannelis C, Gupta A. Does private equity investment in healthcare benefit patients? Evidence from

nursing homes [Internet]. National Bureau of Economic Research 2021 [cited 2022 Nov 15]. (Working Paper Series). Available from: https://www.nber.org/papers/w28474

24. Psychedelic capitalism and the sacred [Internet]. 2021 [cited 2022 Nov 15]. Available from: https://www.youtube.com/watch?v=rf7S0PcfsA4

25. Ltd IAP. Psychedelic therapeutics market worth $ 8.31 billion by 2028 – exclusive report by InsightAce Analytic. [Internet]. [cited 2022 Nov 1]. Available from: https://www.prnewswire.co.uk/news-releases/psychedelic-therapeutics-market-worth-8-31-billion-by-2028-exclusive-report-by-insightace-analytic-816360435.html

26. Londesbrough DJ, Brown C, Northen JS, Moore G, Patil HK, Nichols DE, et al. Treatment of depression and other various disorders with psilocybin [Internet]. Patentscope 2020 [cited 2022 Nov 15]. Available from: https://patentscope.wipo.int/search/en/detail.jsf?docId=WO2020212952&tab=PCTBIBLIO

27. Psychedelic capitalism: debate with Compass Pathways co-Founder [Internet]. 2021 [cited 2022 Nov 15]. Available from: https://www.youtube.com/watch?v=FdOzSHkyLlQ

28. Ibid.

29. Polat O. Interview for The bigger picture. [Personal interview, 2022 Sep] Zoom; 2022 (unpublished).

30. Endgame for the psychedelic renaissance? Jamie Wheal [Internet]. 2021 [cited 2022 Nov 15]. Available from: https://www.youtube.com/watch?v=EOXJOZK1ZNQ

31. Busby M. Inside the Bristol clinic offering ketamine-assisted psychotherapy for mental health and addiction [Internet]. The Bristol Cable 2022 Apr 19 [cited 2022 Dec 14]. Available from: https://thebristolcable.org/2022/04/inside-the-bristol-clinic-trialling-ketamine-assisted-therapy-for-mental-health-and-addiction/

32. Williams M, Negrin D, Eriacho B, Buchanan N, Davis E, de Leon C. Psychedelic justice: toward a diverse and equitable psychedelic culture. Labate BC, Cavnar C, editors. Santa Fe: Synergetic Press; 2021, pp. 5.

33. Stone R. Interview for The bigger picture. [Personal interview, 2022 September] Zoom; 2022 (unpublished).

34. HPPD Nonprofit – The Perception Restoration Foundation [Internet]. Perception Restoration Foundation [cited 2022 Nov 15]. Available from: https://www.perception.foundation

35. Lutkajtis A. The dark side of dharma: Meditation, madness and other maladies on the contemplative path. London: Aeon; 2021.

36. Lutkatjis A. Interview for The bigger picture. [Personal interview, 2022 Aug] Zoom; 2022 (unpublished).

37. Pace BA, Devenot N. Right-wing psychedelia: case studies in cultural plasticity and political pluripotency. Frontiers in Psychology. 2021 Dec 10; 12.

38. Kryskow P. Interview for The bigger picture. [Personal interview, 2022 Oct] Email; 2022 (unpublished).

39. The future of psychedelics with Dennis McKenna [Internet]. 2021 [cited 2022 Nov 15]. Available from: https://www.youtube.com/watch?v=5ae4T90SEg8

40. Amezcua, G. Breaking convention. In: Worlds Collide Conference; 2022 Jul 2; London.

41. Psychedelic capitalism and the sacred [Internet]. 2021 [cited 2022 Nov 15]. Available from: https://www.youtube.com/watch?v=rf7S0PcfsA4

42. Stop and search [Internet]. 2022 May 27 [cited 2022 Nov 15]. Available from: https://www.ethnicity-facts-figures.service.gov.uk/crime-justice-and-the-law/policing/stop-and-search/latest

43. The pipeline. Women count 2021. [Internet]. 2021 [cited 2022 Nov 15] .Available from: https://execpipeline.com/wp-content/uploads/2021/07/Women-Count-2021-Report.pdf

44. Chancel L, Piketty T, Saez E, Zucman G. World inequality report 2022. Cambridge, MA: Harvard University Press; 2022. p. 2.

45. Lorde A. The master's tools will never dismantle the master's house. 1st edition. New York: Penguin Random House; 2018. p. 17.

46. Famurewa J. John Boyega: 'I'm the only cast member whose experience of Star Wars was based on their race.' [Internet]. British

GQ 2020 Sep 2 [cited 2022 Nov 15]. Available from: https://www.gq-magazine.co.uk/culture/article/john-boyega-interview-2020

47. Kim ET. The upstart union challenging Starbucks [Internet]. The New Yorker 2022 Aug 2 [cited 2022 Nov 15]. Available from: https://www.newyorker.com/news/dispatch/the-upstart-union-challenging-starbucks

48. Sorkin AR. Howard Schultz: Starbucks is battling for the 'hearts and minds' of workers. [Internet]. The New York Times 2022 Jun 11 [cited 2022 Nov 15]. Available from: https://www.nytimes.com/2022/06/11/business/dealbook/howard-schultz-starbucks.html

49. Wallace-Wells B. The Marxist who antagonizes liberals and the left. [Internet]. The New Yorker 2022 Jan 21 [cited 2022 Nov 15]. Available from: https://www.newyorker.com/news/annals-of-inquiry/the-marxist-who-antagonizes-liberals-and-the-left

50. Goodhart D. The road to somewhere: the new tribes shaping British politics. UK: Penguin Random House; 2017.

51. Lyons NS. Reality honks back [Internet]. The Upheaval 2022 Feb 16 [cited 2022 Nov 15]. Available from: https://theupheaval.substack.com/p/reality-honks-back

52. Lasch C. The revolt of the elites and the betrayal of democracy. New York: WW Norton; 1996. p. 20.

53. Kendi IX. How to be an Antiracist. 1st edition. New York: Vintage Digital; 2019. p. 234.

54. Pogue J. Inside the new right, where Peter Thiel is placing his biggest bets [Internet]. Vanity Fair 2022 Apr 20 [cited 2022 Nov 15]. Available from: https://www.vanityfair.com/news/2022/04/inside-the-new-right-where-peter-thiel-is-placing-his-biggest-bets

55. Kaschuta A. Interview for The bigger picture. [Personal interview, 2022 Aug] Zoom; 2022 (unpublished).

56. Yarvin C. A brief explanation of the cathedral [Internet]. Gray Mirror 2021 [cited 2022 Nov 15]. Available from: https://graymirror.substack.com/p/a-brief-explanation-of-the-cathedral

57. Meyer MW, Robinson JM. The Nag Hammadi Scriptures: The revised and updated translation of sacred Gnostic texts complete

in one volume. International, reprint, revised, updated edition. San Francisco: HarperOne; 2009.

58. Baudrillard J. Simulacra and Simulation. Ann Arbor: University of Michigan Press; 199. p. 1.

59. Barrett FS, Griffiths RR. Classic hallucinogens and mystical experiences: phenomenology and neural correlates. Current Topics in Behavioral Neurosciences. 2019 August 23.

60. Jung CG. Psychology and alchemy (collected works of C.G. Jung): 12. 2nd edition. London: Routledge; 1980. p. 99.

61. van der Kolk B. Interview for The truth about trauma. [Personal interview, 2022 Apr] Zoom; 2022 (unpublished).

62. Bache CM, Laszlo E. LSD and the mind of the universe: diamonds from heaven. Rochester, VT: Park Street Press; 2020.

63. Langlitz N. Neuropsychedelia: the revival of hallucinogen research since the decade of the brain. 1st edition. Oakland: University of California Press; 2012. p. 204–08.

64. Durkheim É. The elementary forms of religious life. Abridged edition. Cladis MS, editor. Oxford: OUP Oxford; 2008.

65. Lynch G, Sheldon R. The sociology of the sacred: a conversation with Jeffrey Alexander. Culture and Religion: An Interdisciplinary Journal. 2013 Sep 1.

66. Kingsley P. Carl Jung and the end of humanity. 2-volume set. London: Catafalque Press; 2018. p. 23.

67. Ibid, p. 28.

Chapter Five

1. Ep. 22 – Awakening from the meaning crisis – Descartes vs. Hobbes [Internet]. 2019 [cited 2022 Nov 17]. Available from: https://www.youtube.com/watch?v=T-e2Z49n2h8

2. Davis E. Interview for The bigger picture. [Personal interview, 2022 June] In Person; 2022 (unpublished).

3. Edgar Mitchell's strange voyage. [Internet]. People 1974 Apr 8 [cited 2022 Nov 17]. Available from: https://people.com/archive/edgar-mitchells-strange-voyage-vol-1-no-6/

4. Linde, A. Universe, life, consciousness. A paper delivered at the Physics and Cosmology Group of the Science and Spiritual Quest program of the Center for Theology and the Natural Sciences (CTNS); Berkeley, CA; 1998. [Internet]. Available from: web.stanford.edu/~alinde/SpirQuest.doc

5. Langlitz N. Neuropsychedelia: the revival of hallucinogen research since the decade of the brain. 1st edition. Oakland: University of California Press; 2012. p. 207.

6. Kastrup B. Yes, free will exists: just ask Schopenhauer. [Internet]. Scientific American Blog Network 2020 Feb 5 [cited 2022 Nov 17]. Available from: https://blogs.scientificamerican.com/observations/yes-free-will-exists/

7. Kastrup B. Interview for The bigger picture. [Personal interview, 2022 Apr] Zoom; 2022 (unpublished).

8. Kastrup B, Stapp HP, Kafatos M. Coming to grips with the implications of quantum mechanics [Internet]. Scientific American Blog Network 2018 May 29 [cited 2022 Nov 17]. Available from: https://blogs.scientificamerican.com/observations/coming-to-grips-with-the-implications-of-quantum-mechanics/

9. In defence of theology: a reply to Jerry Coyne [Internet]. [cited 2022 Nov 17]. Available from: https://www.bernardokastrup.com/2014/09/in-defence-of-theology-reply-to-jerry.html

10. A conscious universe | Philip Goff [Internet]. IAI TV - Changing how the world thinks 2020 [cited 2022 Nov 17]. Available from: https://iai.tv/articles/conscious-universe-panpsychism-idealism-goff-kastrup-auid-1584

11. Sjöstedt-Hughes P. The white sun of substance: Spinozism and the psychedelic amor dei intellectualis. In: Hauskeller C, Sjöstedt-Hughes P. Philosophy and psychedelics: frameworks for exceptional experience. 1st edition. London/New York: Bloomsbury Academic; 2022. p. 211.

12. McGilchrist I. Interview for The bigger picture. [Personal interview, 2022 Jul] Zoom; 2022 (unpublished).

13. Segall, MD. Altered Consciousness After Descartes: Whitehead's Philosophy of Organism as Psychedelic Realism. In: Hauskeller C, Sjöstedt-Hughes P. Philosophy and Psychedelics: Frameworks

for Exceptional Experience. 1st edition. London/New York: Bloomsbury Academic; 2022. p. 197.

14. Ibid, p. 204.

15. Timmermann C, Kettner H, Letheby C, Roseman L, Rosas FE, Carhart-Harris RL. Psychedelics alter metaphysical beliefs. Sci Rep. 2021 Nov 23; 11(1):22166.

16. Sjöstedt-Hughes P. Interview for The bigger picture. [Personal interview, 2022 July] Zoom; 2022 (unpublished).

17. Kastrup B. Interview for The bigger picture. [Personal interview, 2022 April] Zoom; 2022 (unpublished).

18. Pirsig RM. Lila: Robert M. Pirsig. 1st edition. London: Alma Books; 2011. p. 121.

19. Allan G. The metaphysical axioms and ethics of Charles Hartshorne. The Review of Metaphysics. 1986; 40(2):271–304.

20. Bryant A. Does culture really eat strategy for breakfast? [Internet]. strategy+business 2021 May 27 [cited 2022 Dec 14]; Available from: https://www.strategy-business.com/blog/Does-culture-really-eat-strategy-for-breakfast

21. GameB Home [Internet]. [cited 2022 Nov 17]. Available from: https://www.game-b.org/

22. Humanity's phase shift, Daniel Schmachtenberger [Internet]. 2018 [cited 2022 Nov 17]. Available from: https://www.youtube.com/watch?v=nQRzxEobWco

23. Nguyen CT. Games: agency as art. Oxford: Oxford University Press; 2020. p. 201.

Chapter 6

1. We have to rescue animism: a conversation with Colombian-born anthropologist Luis Eduardo Luna [Internet]. Spirit Territory 2022 Mar 22 [cited 2022 Nov 17]. Available from: https://spiriterritory.com/conversations/interviews/25427-we_have_to_rescue_animism/

2. Nayak SM, Griffiths RR. A single belief-changing psychedelic experience is associated with increased attribution of consciousness to living and non-living entities. Frontiers in

Psychology [Internet]. 2022 Mar 28 [cited 2022 Nov 17]; 13. Available from: https://www.frontiersin.org/articles/10.3389/fpsyg.2022.852248

3. New Johns Hopkins study explores relationship between psychedelics and consciousness [Internet]. Johns Hopkins Medicine Newsroom 2022 Mar 31 [cited 2022 Nov 17]. Available from: https://www.hopkinsmedicine.org/news/newsroom/news-releases/new-johns-hopkins-study-explores-relationship-between-psychedelics-and-consciousness

4. Pollan M. Interview for The bigger picture. [Personal interview, 2022 May] Zoom; 2022 (unpublished).

5. We have to rescue animism: a conversation with Colombian-born anthropologist Luis Eduardo Luna [Internet]. Spirit Territory 2022 Mar 22 [cited 2022 Nov 17]. Available from: https://spiriterritory.com/conversations/interviews/25427-we_have_to_rescue_animism/

6. Stamets P. Mycelium running: how mushrooms can help save the world. Illustrated edition. Berkeley, CA: Ten Speed Press; 2005. p. 4.

7. New Zealand river's personhood status offers hope to Māori. [Internet]. The Independent 2022 Aug 15 [cited 2022 Nov 17]. Available from: https://www.independent.co.uk/news/ap-new-zealand-albert-britain-indiana-jones-b2145077.html

8. Kolbert E. A lake in Florida suing to protect itself. [Internet]. The New Yorker 2022 Apr 11 [cited 2022 Dec 14]. Available from: https://www.newyorker.com/magazine/2022/04/18/a-lake-in-florida-suing-to-protect-itself

9. Pollan M. Interview for The bigger picture. [Personal interview, 2022 May] Zoom; 2022 (unpublished).

10. Will magic mushrooms be legalised? with David Nutt, Crispin Blunt & Tom Eckert [Internet]. 2021 [cited 2022 Nov 17]. Available from: https://www.youtube.com/watch?v=3n5rNMw7C90

11. Leary T, Robbins T, Sirius RU. The politics of ecstasy. 4th edition. Berkeley, CA: Ronin Publishing; 1998. p. 69.

12. Davis T. Interview for The bigger picture. [Personal interview, 2022 August] Zoom; 2022 (unpublished).

13. Beiner A. Face to faith. [Internet]. The Guardian 2009 Oct 9 [cited 2022 Nov 17]; Available from: https://www.theguardian.com/commentisfree/belief/2009/oct/10/drug-use-spiritual-practice

14. The rise of psychedelic capitalism [Internet]. 2021 [cited 2022 Nov 17]. Available from: https://www.youtube.com/watch?v=uXoDOGkmZsI

15. Beiner A. Who's in charge of psilocybin? [Internet]. Chacruna 2021 May 10 [cited 2022 Nov 17]. Available from: https://chacruna.net/who_owes_psilocybin/

16. Clerx A. The rise of evangelical churches. [Internet]. European Academy on Religion and Society 2020 Oct 6 [cited 2022 Nov 18]. Available from: https://europeanacademyofreligionandsociety.com/news/the-rise-of-evangelical-churches/

17. Hofmann A. From molecules to mystery: psychedelic science, the natural world, and beyond. In: Walsh R, Grob CS, editors. Higher wisdom: eminent elders explore the continuing impact of psychedelics. Unabridged edition. New York: State University of New York Press; 2005. p. 51.

18. About Ligare [Internet]. Ligare [cited 2022 Nov 18]. Available from: https://www.ligare.org/about

19. Rosenblum C. These Mormons have found a new faith – in magic mushrooms. [Internet]. Rolling Stone 2022 Jun 28 [cited 2022 Nov 18]. Available from: https://www.rollingstone.com/culture/culture-features/psychedelics-mormon-church-divine-assembly-1375027/

20. Gurdjieff GI. Beelzebub's tales to his grandson: all and everything, first series (book two, enlarged print): an objectively impartial criticism of the life of man. Independently published; 2021.

21. Griffiths RR, Johnson MW, Carducci MA, Umbricht A, Richards WA, Richards BD, et al. Psilocybin produces substantial and sustained decreases in depression and anxiety in patients with life-threatening cancer: a randomized double-blind trial. Journal of Psychopharmacology. 2016 Dec 30; (12):1181–97.

22. Hallucinogenic drug psilocybin eases existential anxiety in people with life-threatening cancer [Internet]. Johns Hopkins Medicine 2016 Dec 2 [cited 2022 Nov 18]. Available from: https://www.hopkinsmedicine.org/news/media/releases/hallucinogenic_

drug_psilocybin_eases_existential_anxiety_in_people_with_
life_threatening_cancer

23. Macdonald L. What psilocybin taught me about living and dying [Internet]. Drug Science 2021 Jan 29 [cited 2022 Nov 18]. Available from: https://www.drugscience.org.uk/psilocybin-living-and-dying/

24. Cooper A. Psilocybin sessions: Psychedelics could help people with addiction and anxiety. CBS News 2019 Oct 13 [cited 2022 Nov 18]. Available from: https://www.cbsnews.com/news/psychedelic-drugs-lsd-active-agent-in-magic-mushrooms-to-treat-addiction-depression-anxiety-60-minutes-2019-10-13/

25. Kastrup B. Interview for The bigger picture. [Personal interview, 2022 April] Zoom; 2022 (unpublished).

26. Kingsley P. Reality. New and updated edition. London: Catafalque Press; 2020.

Conclusion

1. Smith H. Do drugs have religious import? A forty year follow up. In: Walsh R, Grob CS, editors. Higher wisdom: eminent elders explore the continuing impact of psychedelics. Unabridged edition. New York: State University of New York Press; 2005. p. 233.

Alexander Beiner

About the Author

Alexander Beiner is a writer, podcaster, and entrepreneur. He is co-executive director of Breaking Convention, one of Europe's longest-running conferences on psychedelic science and culture. He is also one of the founders of Rebel Wisdom, a media and events platform which has a quarter of a million subscribers. He runs the Substack The Bigger Picture, where he writes about popular culture, sensemaking, systems change, and psychedelics.

As a group facilitator, Alexander has designed and run experiences for thousands of people around the world, both in person and online. He leads and develops legal psychedelic retreats, online courses, and personal development experiences which have been covered by *GQ*, *The Guardian*, and the BBC.

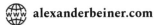 alexanderbeiner.com

 @AlexanderBeiner

 beiner.substack.com

CONNECT WITH

HAY HOUSE

ONLINE

🌐 hayhouse.co.uk   **f** @hayhouse

📷 @hayhouseuk   🐦 @hayhouseuk

▶ @hayhouseuk   ♪ @hayhouseuk

Find out all about our latest books & card decks • Be the first to know about exclusive discounts • Interact with our authors in live broadcasts • Celebrate the cycle of the seasons with us • Watch free videos from your favourite authors • Connect with like-minded souls

'*The gateways to wisdom and knowledge are always open.*'

Louise Hay